Simple quantum physics

Simple quantum physics

PETER LANDSHOFF
Reader in Mathematical Physics, University of Cambridge

ALLEN METHERELL
Lecturer in Physics, University of Cambridge

CAMBRIDGE UNIVERSITY PRESS
CAMBRIDGE
LONDON · NEW YORK · MELBOURNE

Published by the Syndics of the Cambridge University Press
The Pitt Building, Trumpington Street, Cambridge CB2 1RP
Bentley House, 200 Euston Road, London NW1 2DB
32 East 57th Street, New York, NY 10022, USA
296 Beaconsfield Parade, Middle Park, Melbourne 3206, Australia

First published 1979

Printed in Great Britain by
J. W. Arrowsmith Ltd, Bristol

Library of Congress Cataloguing in Publication Data

Landshoff, Peter, 1937–
Simple quantum physics.

Includes index.
1. Quantum theory. I. Metherell, Allen,
1937– joint author. II. Title.
QC174.12.L37 530.1'2 78-73244

ISBN 0 521 22498 5
ISBN 0 521 29538 6 pbk.

CONTENTS

PREFACE

This book is intended as a first course on quantum mechanics and its applications. It is designed to be a first course rather than a complete one, and it is based on lectures given to mathematics and physics students in Cambridge. The book should be suitable also for engineering students.

The first five chapters deal with basic quantum mechanics, and are followed by a revision quiz with which the reader may test his understanding of them. Perhaps the most noticeable omission from these chapters is a detailed discussion of the mathematics of intrinsic spin (Pauli matrices, etc.). Our experience is that students initially find this more difficult than the other material, and since it is not needed for the applications described in the second part of the book, we have omitted it.

In most courses on quantum mechanics, the first application is to scattering problems. While recognising the importance of scattering theory, we have chosen rather to describe the application of quantum mechanics to physical phenomena that are of more everyday interest. These include molecular binding, the physics of masers and lasers, simple properties of crystalline solids arising from their electronic band structure, and the operation of junction transistors.

A few problems are included at the end of each chapter. We urge the student to work through all of these, as they form an integral part of the course. Some hints on their solution may be found at the end of the book.

We express our thanks to Sandra Evans, who typed the manuscript, and to David Branson, Ian Drummond, Sir James Lighthill, Michael Pepper and Ian Smith, who have very kindly read various parts of it and made valuable suggestions.

Christ's College, Cambridge Peter Landshoff
May 1978 Allen Metherell

CONSTANTS OF QUANTUM PHYSICS

Dirac's constant $\hbar = h/2\pi = 1.05 \times 10^{-34}$ J s

Charge of electron $-e = -1.60 \times 10^{-19}$ C

Fine-structure constant $e^2/4\pi\varepsilon_0\hbar c = 1/137$

Speed of light $c = 3.00 \times 10^8$ m s^{-1}

Mass of electron $m_e = 9.11 \times 10^{-31}$ kg

Mass of proton $m_p = 1.67 \times 10^{-27}$ kg

Electron volt 1 eV $= 1.60 \times 10^{-19}$ J

Boltzmann's constant $k_B = 1.38 \times 10^{-23}$ J K^{-1}

1

Preliminaries

Atoms

An atom consists of a positively charged nucleus, together with a number of negatively charged electrons. Inside the nucleus there are protons, each of which carries positive charge e, and neutrons, which have no charge. So the charge on the nucleus is Ze, where Z, the atomic number, is the number of protons. The charge on each electron is $-e$, so that when the atom has Z electrons it is electrically neutral. If some of the electrons are stripped off, the atom then has net positive charge; it has been *ionised*.

The electrons are held in the atom by the electrostatic attraction between each electron and the nucleus. There is also an attraction because of the gravitational force, but this is about 10^{-40} times less strong, and so may be neglected. The protons and neutrons are held together in the nucleus by a different type of force, the nuclear force. The nuclear force is much stronger than the electrical force, and its attraction more than counteracts the electrostatic repulsion between pairs of protons. The nuclear force does not affect electrons. It is a very short-range force, so that it keeps the neutrons and protons very close together; the diameter of a nucleus is of the order of 10^{-15} m. By contrast, the diameter of the whole atom is about 10^{-10} m, so that for many purposes one can think of the nucleus as a point charge. The mass of the proton or neutron is some 2000 times that of the electron, so nearly all the mass of the atom is in the nucleus.

It is natural to think of the electrons as being in orbit round the nucleus, figure 1.1, just as the planets are in orbit round the sun. The electrostatic force that keeps the electrons in their orbits is an inverse-square-law force, just as is the gravitational force that keeps the planets in orbit, so that the two systems would seem to obey precisely similar equations. However, there is a serious difficulty. When a particle moves in a curved orbit its velocity vector is continuously changing: the particle is being accelerated towards the centre of its orbit. According to classical electrodynamics, when a charged particle is accelerated it inevitably radiates energy (this is the basic principle of radio transmission). So according to classical physics the electron would continuously lose energy and its orbit would form a spiral which would gradually collapse into the nucleus.

The reason that this does not happen is that very small systems, such as atoms, do not obey classical mechanics. To describe an atom one has to use quantum mechanics. In quantum mechanics, as opposed to classical mechanics, one cannot arbitrarily choose a value for the energy of the orbiting particle and then find an orbit corresponding to that energy; only certain discrete values of the energy are allowed. When the electron is in its lowest allowed energy level, it cannot radiate any more energy, and so the total collapse of the atom is not possible.

One can also use quantum mechanics to describe the solar system. Just as for the electrons, the allowed energy levels of the planets are discrete. If a planet in orbit is given an impulse, its energy is allowed to change only to that of one of the other allowed discrete levels. However, the separation between these levels is so small that this is not a very real restriction, and classical mechanics is perfectly adequate to describe the system. The effects of quantum mechanics are generally only important for submicroscopic systems.

The chemistry of an atom is determined by the charge on its nucleus. Thus atoms whose nuclei differ only in the number of neutrons that

Figure 1.1. Classical picture of negatively charged electrons in orbit round the positively charged nucleus of an atom.

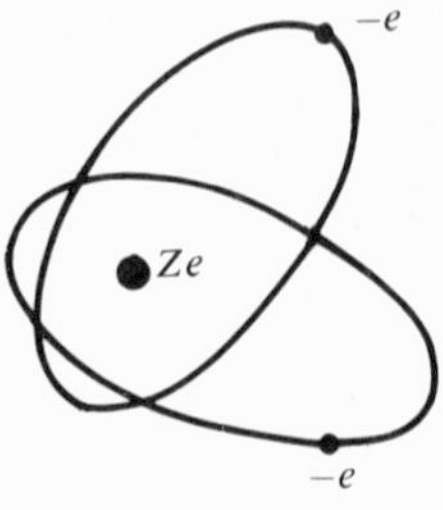

they contain have similar chemical properties; they are said to be *isotopes* of the same element. For example, the atom of the common form of hydrogen contains just a single proton, that is, $Z=1$; but hydrogen has a stable isotope, called deuterium, whose nucleus consists of one proton and one neutron. Atoms can be bound together to form molecules (see chapter 6), and different isotopes of the same element do this in the same way. Ordinary water consists of molecules containing two hydrogen nuclei and one oxygen nucleus, H_2O, while 'heavy' water has deuterium nuclei instead of the ordinary hydrogen, D_2O. The chemical properties of heavy water are exactly the same as those of ordinary water, but there are some differences in its physical properties. In particular, it is denser because of the extra neutrons.

Photons

In a metal, the atoms are effectively anchored to fixed sites by the electrostatic forces due to all the other atoms. The outermost orbital electrons of the atoms are almost free, and move through the metal when an electric field is applied. (See chapter 11.) If the metal is bombarded with light, some of the electrons can actually escape from the surface of the metal and can be detected as an electric current. This is the *photoelectric effect.* The number of electrons that escape in a given time rises with the intensity of the beam of light, but the energy with which they escape does not depend on the beam intensity. Rather it depends on the colour or frequency ν of the light. The kinetic energy T with which the electrons escape is found to obey the equation

$$h\nu = T + W. \tag{1.1}$$

Here h is Planck's constant,

$$h = 6.626 \times 10^{-34}\ \mathrm{J\ s},$$

and W is the energy that must be given to the electron to enable it to overcome the electrostatic attraction of the metal. (The value of W varies, according to the state within the metal from which the electron is ejected. For a given metal, there is a definite minimum value W_0, called the *work function* of the metal.)

These results are explained by the hypothesis that a beam of light can be thought of as a collection of particles, called *photons.* The number of photons is proportional to the intensity of the light, and the energy E of each photon is proportional to the frequency,

$$E = h\nu. \tag{1.2}$$

The electron is ejected from the metal when one of the photons collides with it and is absorbed by it, so giving up all its energy to the electron. The number of electrons ejected rises as the intensity of the light is increased because there are then more incident photons, and so there is a greater chance of a photon being absorbed.

Photons move with the speed of light, so their kinematics must be described by the laws of special relativity. The energy of a particle whose speed is v and whose rest mass is m is

$$E = mc^2/(1 - v^2/c^2)^{1/2}, \tag{1.3a}$$

so that when $v = c$ the energy can be finite only if $m = 0$; that is, photons have zero mass. In terms of the momentum p of the particle, (1.3a) reads

$$E = c(m^2c^2 + p^2)^{1/2}, \tag{1.3b}$$

so that for a photon

$$E = cp. \tag{1.4}$$

If a beam of light is shone normally on a perfect conductor it is reflected, that is, the momentum of each photon is reversed. This must occur by some sort of force being exerted on the photons, and the conductor must experience an equal and opposite force. This is a realisation of the classical idea of *radiation pressure.*

The equations (1.2) and (1.4) are tested in the *Compton effect.* When photons collide with free electrons or protons, not bound into a solid, they cannot be absorbed because it can be shown (see problem 1.2) that this would violate conservation of energy and momentum. (In the photoelectric effect some of the energy, W, is absorbed by the other particles in the metal). However, a free particle can scatter the photon, so changing its energy and therefore its frequency; at the same time the particle recoils. The kinematics of the process can be worked out using (1.2), (1.4) and the relativistic energy–momentum conservation laws (see problem 1.4), and the results are found to agree with experiment.

The equation (1.2) also helps to explain atomic spectra. We have said that, according to the results of quantum mechanics, the allowed energy levels of the electrons in atoms are discrete. If a beam of light is shone on a collection of atoms, the photons can be absorbed by the atoms if, and only if, their energy is equal to the difference between the energies of two electron levels. The absorption of the photon then excites the atom, sending the electron from the lower to the higher level. (The number of photons that can be absorbed in this way of course depends on how many atoms happen to be in the lower of the two states to start with.) Thus only photons with certain discrete frequencies are absorbed. Conversely, an atom in an excited state can decay by emitting a photon; the frequency of the photon depends on the difference between the initial and final energies:

$$h\nu = E_2 - E_1.$$

The energy levels of an atom (or molecule) depend on what element it is, so that the spectrum of frequencies absorbed and emitted provides a useful way of identifying substances.

Wave nature of matter

Although light can be thought of as a collection of photons, it also has wave-like properties. For example, a coherent beam of light is diffracted when it is shone through a pair of closely separated slits: if a screen is placed at large distance behind the slits, a pattern of light and dark fringes appears on it. The spacing of these fringes is calculated from the wavelength λ of the light. See figure 1.2. Dark fringes appear at points on the screen such that their distances from the two slits differ by $(n+\frac{1}{2})\lambda$, where n is an integer. Then the light received from the two slits is exactly out of phase; the two components cancel.

So quantum mechanics gives light a dual nature. In some respects it behaves like a collection of particles, in others like a wave. The same is

Figure 1.2. The double-slit experiment. There is darkness at points on the screen such that the path difference between rays that pass through the two slits is $(n+\frac{1}{2})\lambda$.

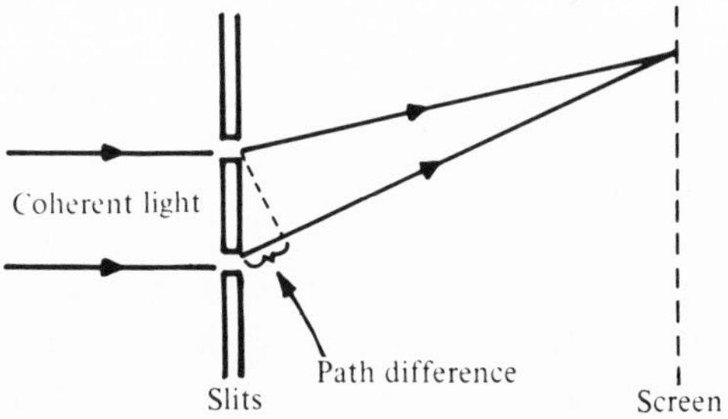

true for electrons and other particles; quantum mechanics associates waves with every kind of particle.

The necessity for this is illustrated by the phenomenon of electron diffraction. If a beam of electrons is passed through a crystal, it is diffracted. If a fluorescent screen is set up behind the crystal, a diffraction pattern appears on the screen. The regularly spaced atoms in the crystal cause the diffraction. The pattern can be explained by associating with the electrons a wave of wavelength λ, which changes with the momentum of the electrons according to de Broglie's relation

$$\lambda = h/p. \tag{1.5}$$

The waves may be assigned a characteristic frequency ν, which may be chosen so as to be related to the electron energy E by the relation $E = h\nu$ as for photons (see (1.2)). However, in the case of electrons, the frequency ν is not directly measurable and there is some arbitrariness in its definition; for example, E may or may not include the rest-mass energy of the electron. The part played in the theory by the electron frequency will become apparent in the next chapter.

The de Broglie relation (1.5) applies also to photons. This follows if we make the assumption that the quantum-mechanical waves that describe photons have the same frequency ν and wavelength λ as the corresponding classical electromagnetic waves. Because the classical waves have speed c, this implies that

$$\lambda\nu = c \tag{1.6}$$

and combining this relation with (1.2) and (1.4) gives (1.5).

Classical electromagnetic waves are associated with a very large number of photons (see problem 1.1). The waves of quantum mechanics may describe either a collection of particles or a single particle. It is important to understand that quantum-mechanical waves are more abstract than classical waves. Consider an experiment where light is diffracted through a pair of slits, or where electrons are diffracted through a crystal. Suppose that only one photon or electron is allowed to come into the experiment. In this case we cannot predict with certainty what will be the angle θ through which the photon or electron is diffracted. But if the experiment is repeated many times, we find a *probability distribution* for the angle θ that has the same shape as the variation of intensity with θ in an experiment where there is a continuous beam of photons or electrons.

This suggests that the association of a quantum-mechanical wave with a photon, or with any other kind of particle, is somehow statistical. We explore this in the following chapters. As will become clear, according to quantum theory one can never predict with certainty what will be the result of a particular experiment: the best that can be done is to calculate the *probability* that the experiment will have a given result, or one can calculate the *average value* of an observable quantity if the experiment that measures it is repeated many times.

Problems

1.1 A radio transmitter operates on a wavelength of 100 m at a power of 1 kW. How many photons does it emit per second?

1.2 Using energy–momentum conservation, show that an electron that is not in a bound state cannot absorb a photon.

1.3 A particle has mass 1 kg. How long does it take to move through a distance of 1 m if its de Broglie wavelength, (1.5), is comparable with the wavelength of visible light? What is the corresponding answer if the particle is an electron?

1.4 A photon of momentum p, and therefore of wavelength h/p, scatters on an electron that is initially at rest. Using relativistic kinematics, deduce from the conservation of energy and momentum that as the result of the scattering the wavelength of the photon changes by $(h/mc)(1-\cos\theta)$, where θ is the angle through which it scatters and m is its rest mass. (This scattering process is known as *Compton scattering*, and the quantity h/mc is the *Compton wavelength* of the electron.)

1.5 Associated with the electron there is an *antiparticle*, the positron, which has equal mass and equal, but opposite, charge.

A positron impinges on an electron which is at rest. They annihilate into two photons. Show that the sum of the wavelengths of the two photons is $\lambda_0(1-\cos\theta)$, where θ is the angle between their directions of motion and λ_0 is the Compton wavelength of the electron.

2

The Schrödinger equation

The basic equation of quantum mechanics is known as the Schrödinger equation. It is not possible to prove that the equation must be true, just as we cannot prove that Newton's laws, the basis of classical mechanics, must be true. All that can be done is to work out the consequences of the equation in different physical contexts, and to compare them as exhaustively as possible with experiment. In this chapter we begin by showing how the experimental facts that we have already described make the truth of the equation *plausible*.

The Schrödinger equation describes non-relativistic particles, whose energy E and momentum p are related by

$$E = p^2/2m. \tag{2.1}$$

Non-relativistic kinematics can be used so long as the energy E is not comparable with, or larger than, the rest-mass energy mc^2. Relativistic quantum mechanics is much more difficult than the non-relativistic theory, and will not concern us in this book. In particular, we shall not describe the quantum mechanics of the photon; for this particle $m = 0$, and so relativistic mechanics must always be used.

Wave functions and operators

We explained in chapter 1 that with a particle of energy E and momentum p we somehow associate a wave of frequency $\nu = E/h$ and wavelength $\lambda = h/p$. Instead of using ν and λ, it is convenient to introduce the angular frequency

$$\omega = 2\pi\nu$$

and the wave vector $\boldsymbol{k}$, whose direction is in the direction of wave propagation and whose magnitude is

$$k = 2\pi/\lambda.$$

We also work with

$$\hbar = h/2\pi$$

(pronounced 'h-cross') instead of h. Then we have

$$\begin{aligned} \boldsymbol{p} &= \hbar \boldsymbol{k} \\ E &= \hbar\omega. \end{aligned} \tag{2.2}$$

For a free particle, which is not interacting with any other particle or with a potential, $\boldsymbol{p}$ and E are constant. Hence we expect such a particle to be described by a wave for which $\boldsymbol{k}$ and ω are constant:

$$\Psi(\boldsymbol{r}, t) = N\,\mathrm{e}^{\mathrm{i}(\boldsymbol{k}\cdot\boldsymbol{r}-\omega t)}. \tag{2.3}$$

Here $\boldsymbol{r}$ denotes the position vector and N is a constant. In many physical applications, we write a wave in the form of a complex exponential, as is done here, and understand that only the real part is physically meaningful. In quantum mechanics, however, it will turn out that both the real and the imaginary parts of Ψ are needed. The plane wave (2.3) is the simplest example of a *wave function*: it describes a free particle. Instead of (2.3), we could use $N\,\mathrm{e}^{-\mathrm{i}(\boldsymbol{k}\cdot\boldsymbol{r}-\omega t)}$; the choice of sign is a matter of convention.

In order to guess how to arrive at wave functions that describe particles that are not free, we perform a simple manipulation on (2.3). Differentiation with respect to a component x_j ($j = 1$, 2 or 3) of $\boldsymbol{r}$ simply multiplies Ψ by i times the corresponding component k_j of $\boldsymbol{k}$. So, with (2.2), we have

$$(-\mathrm{i}\hbar\,\partial/\partial x_j)\Psi = p_j\Psi \qquad (j = 1, 2, 3) \tag{2.4a}$$

or, in more concise vector notation,

$$(-\mathrm{i}\hbar\boldsymbol{\nabla})\Psi = \boldsymbol{p}\Psi. \tag{2.4b}$$

The equation (2.4a) says that if we apply the differential operator $-\mathrm{i}\hbar\,\partial/\partial x_j$ to Ψ, the result is to multiply Ψ by the number p_j. We say that Ψ is an *eigenfunction* of the operator $(-\mathrm{i}\hbar\,\partial/\partial x_j)$ with *eigenvalue* p_j. (More properly, since the differential equations (2.4) do not determine Ψ until we have imposed suitable boundary and continuity conditions on Ψ, we should at this stage specify these conditions. However, we defer the discussion of this until the end of this chapter.) In this way, an experimentally observable quantity, the momentum $\boldsymbol{p}$ of a particle, is associated with a differential operator $-i\hbar\boldsymbol{\nabla}$. We generalise this to other experimentally observable quantities by introducing three *basic assumptions*:

(*a*) States of a system are described by wave functions Ψ.
(*b*) Observable quantities are associated with operators.
(*c*) When the value of an observable Q is known to be q, the system is in a state whose wave function is an eigenfunction of the operator $\hat{Q}$ corresponding to Q, with eigenvalue q. That is,

$$\hat{Q}\Psi = q\Psi. \tag{2.5}$$

We shall elaborate on these assumptions in chapter 4.

Consider the particular example of the observable that is the energy of a particle. If it is a free particle, this is just $\boldsymbol{p}^2/2m$. So one would expect the corresponding differential operator to be obtained by replacing $\boldsymbol{p}$ in this expression by its corresponding operator $-i\hbar\boldsymbol{\nabla}$, and so we have $(-i\hbar\boldsymbol{\nabla})^2/2m$, that is, $-\hbar^2\nabla^2/2m$. If the particle moves in a potential $V(\boldsymbol{r})$, this operator becomes†

$$H = (-\hbar^2/2m)\nabla^2 + V(\boldsymbol{r}) \tag{2.6}$$

corresponding to the classical expression $p^2/2m + V(\boldsymbol{r})$. H is called the Hamiltonian operator. To find the possible energy levels E of the particle, we must find the eigenvalues E of H, that is, we must solve the equation

$$H\Psi = E\Psi \tag{2.7a}$$

(subject to appropriate boundary conditions). This equation

$$[(-\hbar^2/2m)\nabla^2 + V(\boldsymbol{r})]\Psi(\boldsymbol{r}, t) = E\Psi(\boldsymbol{r}, t) \tag{2.7b}$$

is the *time-independent Schrödinger equation.*

Notice that if the particle is not a free particle, so that $V \neq 0$, the plane wave (2.3) is not a solution of the Schrödinger equation; the wave function Ψ will be more complicated.

The time-independent Schrödinger equation applies when the particle is in a state such that its energy takes a definite value E. In a general situation this may not be the case, for example when V depends explicitly on the time t as well as on the position $\boldsymbol{r}$ of the particle. In order to guess what is the generalisation of the Schrödinger

† Although H is an operator, we shall follow the usual convention and not write the operator symbol ˆ over it.

equation that describes this, we return again to the plane-wave solution (2.3) appropriate to a free particle. Making use now also of the second relation in (2.2), we find that the plane wave satisfies

$$(-\hbar^2/2m)\nabla^2\Psi = \mathrm{i}\hbar(\partial/\partial t)\Psi.$$

The obvious generalisation of this to the case where $V \neq 0$ is

$$[(-\hbar^2/2m)\nabla^2 + V]\Psi(\boldsymbol{r}, t) = \mathrm{i}\hbar\, \partial\Psi/\partial t \tag{2.8a}$$

or

$$H\Psi = \mathrm{i}\hbar\, \partial\Psi/\partial t. \tag{2.8b}$$

This is the *time-dependent Schrödinger equation.*

This is the equation that is valid in all circumstances whether or not the particle is free and whether or not it is known to be in a state of definite energy. In the particular case where the particle is known to be in a state of definite energy E, the time-independent equation (2.7) is valid also. When both equations hold,

$$\mathrm{i}\hbar\, \partial\Psi/\partial t = E\Psi.$$

This means that the time dependence of $\Psi(\boldsymbol{r}, t)$ is then trivial:

$$\Psi(\boldsymbol{r}, t) = \psi(\boldsymbol{r})\mathrm{e}^{-\mathrm{i}Et/\hbar}. \tag{2.9}$$

The function $\psi(\boldsymbol{r})$ is called the *time-independent wave function* and it satisfies the time-independent Schrödinger equation

$$H\psi = E\psi. \tag{2.7c}$$

We repeat that a solution of the form (2.9) is applicable only when the energy takes a definite value. For reasons that will be explained below, a solution of this type is known as a *stationary-state* solution. Comparing (2.9) with (2.3), we see that in the special case of a free particle

$$\psi(\boldsymbol{r}) = N\, \mathrm{e}^{\mathrm{i}\mathbf{k}\cdot\mathbf{r}} \tag{2.10a}$$

where, from (2.7),

$$E = \hbar^2 k^2/2m. \tag{2.10b}$$

Example: the one-dimensional potential well

It turns out that for a particle in a bound state, for example an electron bound in an atom, the values of E that are allowed by the time-independent Schrödinger equation are discrete. That is, the energy levels of a bound state are quantised.

The simplest mathematical model of a bound-state situation is as follows. Suppose that the wave function ψ depends on only one coordinate, x say, and let the potential be the *infinite square well*

$$V = 0 \qquad 0 < x < a$$
$$= \infty \qquad \text{otherwise.} \tag{2.11}$$

As only one of the three coordinates is explicitly involved, we call this a one-dimensional model. To find the states of definite energy E we solve the time-independent Schrödinger equation, which here reduces to

$$[(-\hbar^2/2m)(\mathrm{d}^2/\mathrm{d}x^2) + V]\psi(x) = E\psi(x). \tag{2.12}$$

This equation is to be valid for all x. If E is finite, we deduce that in the region outside $0 < x < a$, where V is infinite, the wave function $\psi(x)$ must vanish. Otherwise the term $V\psi$ in the Schrödinger equation (2.12) would be infinite in this region, which is not possible because neither of the other two terms in the equation would be similarly infinite, and so the equation would not be satisfied. In the region $0 < x < a$, where $V = 0$, the Schrödinger equation reduces to

$$(-\hbar^2/2m)\psi''(x) = E\psi(x)$$

and its general solution for ψ is a linear combination of sin kx and cos kx, where $E = \hbar^2 k^2/2m$.† As we shall see at the end of this chapter, $\psi(x)$ has to be continuous; in particular, it has to be continuous at $x = 0$ and $x = a$. Hence, since it vanishes for $x < 0$ and for $x > a$, we must impose the conditions $\psi(0) = \psi(a) = 0$. The condition at $x = 0$ picks out the sin kx solution for ψ, and that at $x = a$ imposes a restriction on the allowed values of k:

$$k = n\pi/a \qquad n = 1, 2, \ldots$$

We could also allow negative integers n, but this would not give any additional wave functions; the allowed wave functions are

$$\psi_n(x) = C_n \sin(n\pi x/a) \qquad 0 < x < a$$
$$= 0 \qquad \text{elsewhere} \tag{2.13a}$$

where the C_n are normalising constants (we explain below how these are chosen). The corresponding allowed energy values are

$$E_n = n^2 \hbar^2 \pi^2/2ma^2 \qquad n = 1, 2, \ldots \tag{2.13b}$$

That is, the allowed energy values for the bound system are discrete.

† Throughout we shall use primes to denote differentiation with respect to a space variable and dots to denote differentiation with respect to a time variable.

Probability interpretation and normalisation

As we have said in chapter 1, the wave function Ψ associated with a particle in quantum mechanics does not describe a concrete wave; it is not like the waves of classical physics. Indeed, it was not until after it had been discovered that the Schrödinger equation gives the right energy levels of a system that the correct interpretation of Ψ was suggested. We explained that the wave function has a statistical interpretation, and the *basic assumption* is that $|\Psi(\boldsymbol{r}, t)|^2$ is a probability density. That is, $|\Psi(\boldsymbol{r}, t)|^2 \, d^3\boldsymbol{r}$ is the probability that a measurement of the position of the particle described by Ψ will give a result in the infinitesimal volume element $d^3\boldsymbol{r}$ in the neighbourhood of the point $\boldsymbol{r}$. In order for this interpretation to be consistent, Ψ must be arranged to satisfy the *normalisation condition*

$$\int d^3\boldsymbol{r} \, |\Psi(\boldsymbol{r}, t)|^2 = 1. \tag{2.14}$$

Here we have used the symbol $\int d^3\boldsymbol{r}$ as a shorthand for $\iiint dx \, dy \, dz$, and the integration is supposed to extend over all space. So (2.14) corresponds to saying that there is unit probability that a measurement of the position of the particle finds that the particle is actually somewhere.

If we have found a solution Ψ to the Schrödinger equation that is not correctly normalised so as to satisfy (2.14), we can usually get one that is by simply multiplying by a suitable constant. An exception is the special case of the plane wave (2.3), for which the normalising integral (2.14) diverges for all values of the multiplying constant N. The reason for this is that strictly the plane wave does not correspond to a physical situation. It has $|\Psi|^2 = |N|^2$, independent of $\boldsymbol{r}$, so that it gives equal probability of finding the particle anywhere throughout space. In practice, one knows that the particle is confined within some volume Ω_0, for example within a given building, and that therefore the wave function vanishes outside Ω_0. No problems arise for finite Ω_0, as then the normalisation integral converges, and by dividing Ψ by a suitable finite constant we can make it converge to 1, as is required in (2.14). The plane-wave solution (2.3) is best thought of as a mathematical limit, where $\Omega_0 \to \infty$, of a physically allowed solution. Apart from the normalisation problem, which can in fact be circumvented by methods that we shall not discuss here, the mathematics of the $\Omega_0 \to \infty$ situation is rather simpler than that of the proper wave functions, and if the volume Ω_0 is large (for example when Ω_0 represents the volume of a typical experimental apparatus, which is almost always very large when measured on a quantum-mechanical scale), the numerical results are changed very little by taking the limit $\Omega_0 \to \infty$.

Suppose that the particle is confined to a finite volume Ω_0, so that its wave function Ψ vanishes outside that volume. Then if $\Psi(\boldsymbol{r}, t)$ is correctly normalised at a given time t, it remains so. For, if we integrate over *any* volume Ω,

$$\frac{\mathrm{d}}{\mathrm{d}t}\int_\Omega \mathrm{d}^3\boldsymbol{r}\,|\Psi|^2 = \int_\Omega \mathrm{d}^3\boldsymbol{r}\,\frac{\partial}{\partial t}(\Psi^*\Psi)$$
$$= \int_\Omega \mathrm{d}^3\boldsymbol{r}(\Psi^*\dot{\Psi} + \dot{\Psi}^*\Psi). \tag{2.15a}$$

To evaluate this, we use the Schrödinger equation and its complex conjugate:

$$[(-\hbar^2/2m)\nabla^2 + V]\Psi = \mathrm{i}\hbar\dot{\Psi}$$
$$[(-\hbar^2/2m)\nabla^2 + V]\Psi^* = -\mathrm{i}\hbar\dot{\Psi}^*.$$

Thus

$$\frac{\mathrm{d}}{\mathrm{d}t}\int_\Omega \mathrm{d}^3\boldsymbol{r}\,|\Psi|^2 = -\frac{\hbar}{2m\mathrm{i}}\int_\Omega \mathrm{d}^3\boldsymbol{r}(\Psi^*\nabla^2\Psi - \Psi\nabla^2\Psi^*)$$
$$= -\frac{\hbar}{2m\mathrm{i}}\int_\Omega \mathrm{d}^3\boldsymbol{r}\,\boldsymbol{\nabla}\cdot(\Psi^*\boldsymbol{\nabla}\Psi - \Psi\boldsymbol{\nabla}\Psi^*). \tag{2.15b}$$

Figure 2.1. The integration volume for (2.15), and its bounding surface S.

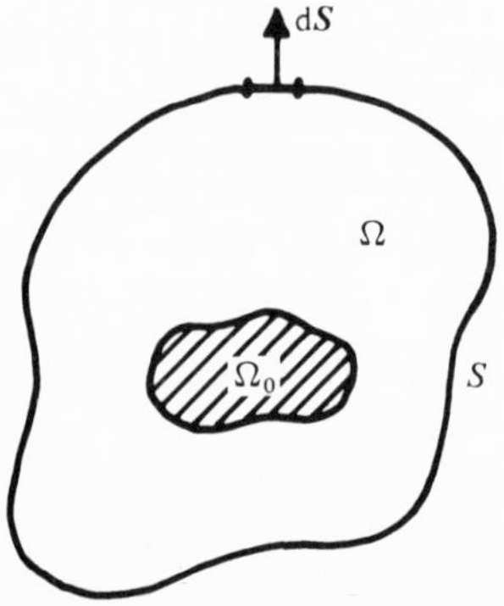

The divergence theorem† can now be used to convert this to an integral over the surface S that encloses the volume of integration (see figure 2.1):

$$\frac{\mathrm{d}}{\mathrm{d}t}\int_{\Omega} \mathrm{d}^3\boldsymbol{r}\, |\Psi|^2 = -\int_S \mathrm{d}\boldsymbol{S}\cdot\boldsymbol{j}$$

where

$$\boldsymbol{j} = (\hbar/2m\mathrm{i})(\Psi^*\boldsymbol{\nabla}\Psi - \Psi\boldsymbol{\nabla}\Psi^*). \tag{2.15c}$$

This relation is valid for any volume Ω. For the normalisation integral (2.14), we have to integrate over all space. But since we are assuming that Ψ vanishes outside some volume Ω_0, it is sufficient to integrate over any volume that contains Ω_0. Because the surface S is then outside the volume Ω_0, Ψ, and therefore $\boldsymbol{j}$, vanishes all over it, and so the integral (2.15) vanishes.

What we have said so far about normalisation applies to any wave function $\Psi(\boldsymbol{r}, t)$. When the wave function corresponds to a state of definite energy, its time dependence factorises out trivially, as in (2.9). In this case

$$|\Psi(\boldsymbol{r}, t)|^2 = |\psi(\boldsymbol{r})|^2$$

so that the probability density at each point $\boldsymbol{r}$ is independent of time. This is the reason that we call a state of definite energy a *stationary state*.

Beams of particles

So far, we have considered a single particle, such as an electron. Experiments often involve beams of particles, for example a beam of electrons from an accelerator fired at an atom. If the beam is not too intense, that is, the particle density is not too high, each electron interacts with the atom independently of the presence of the other electrons, and interactions between the electrons are negligible. This

† The divergence theorem states that for any suitably well-behaved vector $\boldsymbol{u}$

$$\int_{\Omega} \mathrm{d}^3\boldsymbol{r}\,\boldsymbol{\nabla}\cdot\boldsymbol{u} = \int_S \mathrm{d}\boldsymbol{S}\cdot\boldsymbol{u}.$$

Here the vector $\mathrm{d}\boldsymbol{S}$ is a vector whose magnitude is equal to an element $\mathrm{d}S$ of the area of the surface S bounding the volume Ω, and its direction is that of the outward normal to S. See, for example, H. Jeffreys and B. S. Jeffreys, *Methods of Mathematical Physics*, 3rd edn (Cambridge University Press, 1956), p. 193.

means that we may describe the problem in terms of the same Schrödinger equation as describes a single electron moving in the potential produced by the atom.

Suppose that when there is just one electron the potential produced by the atom results in a wave function $\Psi(\boldsymbol{r}, t)$. In the case of an electron beam, the same wave function describes the density of particles at the point $\boldsymbol{r}$ at time t. However, it is convenient then to normalise Ψ not by the integral (2.14) appropriate for a single-particle problem, but instead such that the integral $\int d^3\boldsymbol{r}\, |\Psi(\boldsymbol{r}, t)|^2$ is the total number of particles at time t. Then

$$|\Psi(\boldsymbol{r}, t)|^2 = \rho(\boldsymbol{r}, t). \tag{2.16}$$

Here $\rho(\boldsymbol{r}, t)$ is the average density that would be measured by setting up the same state many times and repeatedly measuring the particle density. This average value is called the *expectation value* of the particle density.

In addition to the expectation value of the particle density, ρ, we may define $\boldsymbol{j}$, the expectation value of the particle flux. This is such that, if we take any closed surface S,

$\int d\boldsymbol{S} \cdot \boldsymbol{j} =$ the expectation value of the number of particles per unit time that cross the surface S.

A positive value for this integral would correspond to a depletion of the total number of particles within S:

$$\frac{d}{dt}\int d^3\boldsymbol{r}\,\rho(\boldsymbol{r}, t) = -\int d\boldsymbol{S} \cdot \boldsymbol{j}. \tag{2.17}$$

With Ψ now normalised according to (2.16), this is just the relation (2.15c). That is, the appropriate expression for $\boldsymbol{j}$ is

$$\boldsymbol{j} = \frac{\hbar}{2m\mathrm{i}}(\Psi^*\nabla\Psi - \Psi\nabla\Psi^*). \tag{2.18}$$

In most cases, the potential and the rate at which the beam delivers particles to the target vary very slowly, if at all, on the atomic timescale. Hence there is effectively a stationary-state situation, with the wave function factorising as in (2.9) and ρ and $\boldsymbol{j}$ independent of t.

Continuity conditions

The Schrödinger equation (2.8) is supposed to be valid at all points $\boldsymbol{r}$. This means that if the potential V is well behaved, the wave function and its space derivative must both be continuous functions of $\boldsymbol{r}$. For if this were not so, the second derivative $\nabla^2\Psi$ would be infinite at the points of discontinuity, and this cannot be because there cannot be just one term in the Schödinger equation that has an infinity. The infinity would have to be balanced by another infinite term and there is no such term unless V has an infinity.

The physical consequences of this are not surprising: both $\rho(\boldsymbol{r}, t)$ and $\boldsymbol{j}(\boldsymbol{r}, t)$ have no discontinuous changes as $\boldsymbol{r}$ varies. That is, no particles are being created or destroyed at any point $\boldsymbol{r}$.

Problems

2.1 How are the stationary-state solutions of the Schrödinger equation changed if a constant is added to the potential $V(\boldsymbol{r})$? Show that this change has no observable consequences. (This means that, as in classical mechanics, the point at which V vanishes may be chosen arbitrarily. By convention, V is often defined so as to vanish at infinity.)

2.2 For each of the following, estimate the difference between the speeds of the particle when it is in the first excited state and when it is in the lowest-energy state: (i) an electron confined by an infinite square-well potential whose width is roughly equal to the radius of an atom (about 10^{-10} m); (ii) a tennis ball confined by an infinite square-well potential whose width is equal to the width of a tennis court. What do your answers tell you about the mechanics of microscopic and macroscopic systems?

2.3 Determine the constant C_n in the wave function (2.13a) for the infinite square well, so as to satisfy the normalisation condition (2.14). Show that the probability interpretation of the wave function leads to the average values

$$\langle x\rangle = \tfrac{1}{2}a, \qquad \langle (x-\langle x\rangle)^2\rangle = (a^2/12)(1-6/n^2\pi^2).$$

Show that as $n\to\infty$ these average values approach the values that are obtained from classical mechanics.

2.4 Show that for a one-dimensional system in a stationary state the particle flux (see (2.18)) is independent of both t and x. Does this result have a simple generalisation to the three-dimensional case?

2.5 A system consists of two particles, so that its stationary-state wave functions $\psi(\boldsymbol{r}_1, \boldsymbol{r}_2)$ are functions of the coordinates of each. What is the obvious probability interpretation of ψ?

Write down the time-independent Schrödinger equation. In the general case, the potential will be a function $V(\boldsymbol{r}_1, \boldsymbol{r}_2)$. What is the structure of the function V in the special case where the particles interact only with an external field of force, and not with each other? Show that in this case the stationary-state wave functions have the structure $\psi(\boldsymbol{r}_1, \boldsymbol{r}_2) = \psi_1(\boldsymbol{r}_1)\psi_2(\boldsymbol{r}_2)$. What equations do the functions ψ_1 and ψ_2 obey?

3

Special solutions

Particle in a box

Consider the stationary states of a particle confined within the rectangular-box-shaped volume

$$0 < x < a, \qquad 0 < y < b, \qquad 0 < z < c. \tag{3.1}$$

Suppose that within the box the potential is zero and outside it is infinite. Within the box, the particle moves freely and the time-independent Schrödinger equation is

$$-\frac{\hbar^2}{2m}\left(\frac{\partial^2}{\partial x^2}+\frac{\partial^2}{\partial y^2}+\frac{\partial^2}{\partial z^2}\right)\psi(\boldsymbol{r}) = E\psi(\boldsymbol{r}). \tag{3.2}$$

Elementary solutions to this equation are

$$\psi(\boldsymbol{r}) = {}^{\sin}_{\cos}\, k_1 x \; {}^{\sin}_{\cos}\, k_2 y \; {}^{\sin}_{\cos}\, k_3 z$$

where

$$E = (\hbar^2/2m)(k_1^2 + k_2^2 + k_3^2). \tag{3.3}$$

As we explained for the example of the one-dimensional infinite square-well potential in chapter 2, because the potential is infinite at all points outside the box, the wave function $\psi(\boldsymbol{r})$ must vanish for $\boldsymbol{r}$ outside the box. To make $\psi(\boldsymbol{r})$ continuous, it must also vanish on the faces of the box. Hence (3.3) must vanish on each of the six planes

$$x = 0, \; y = 0, \; z = 0$$

$$x = a, \; y = b, \; z = c.$$

The first three conditions pick out the sin functions, rather than the cos functions. The last three conditions impose restrictions on the allowed values of k_1, k_2 and k_3:

$$k_1 = \frac{q\pi}{a}, \qquad k_2 = \frac{r\pi}{b}, \qquad k_3 = \frac{s\pi}{c},$$

where q, r and s are integers. Thus the allowed wave functions are

$$\psi_{qrs}(\boldsymbol{r}) = \left(\frac{8}{abc}\right)^{1/2} \sin\frac{q\pi x}{a} \sin\frac{r\pi y}{b} \sin\frac{s\pi z}{c}, \tag{3.4}$$

and the allowed values of the energy are

$$E_{qrs} = \frac{\hbar^2\pi^2}{2m}\left(\frac{q^2}{a^2}+\frac{r^2}{b^2}+\frac{s^2}{c^2}\right). \tag{3.5}$$

Notice that all these allowed values of E_{qrs} are positive. In (3.4) we have included the normalisation factor $(8/abc)^{1/2}$, so as to satisfy the normalisation condition

$$\int \mathrm{d}^3\boldsymbol{r}\, |\psi_{qrs}(\boldsymbol{r})|^2 = 1,$$

the integration being over the volume of the box, or equivalently, since $\psi = 0$ outside the box, over all space.

The significant feature of these results is that the allowed values E_{qrs} of the energy are discrete, because of the condition that q, r and s are integers. Notice that we may as well confine them to being non-negative integers, since changing the sign of any one of them merely changes the sign of the wave function, a change that has no physical significance.

These are the stationary-state solutions. The corresponding time-varying wave functions are, from (2.9),

$$\Psi_{qrs}(\boldsymbol{r}, t) = \psi_{qrs}(\boldsymbol{r})\, \mathrm{e}^{-\mathrm{i}E_{qrs}t/\hbar}. \tag{3.6}$$

In general, if we imagine that we somehow introduce the particle into the box at time $t = 0$, it is unlikely that we will succeed in putting it into a stationary state. By superposing the stationary-state solutions (3.6),

$$\Psi(\boldsymbol{r}, t) = \sum_{q,r,s=1}^{\infty} c_{qrs}\Psi_{qrs}(\boldsymbol{r}, t), \tag{3.7}$$

where the c_{qrs} are constants, we obtain more general wave functions Ψ, which can be verified to satisfy the time-dependent Schrödinger equation (2.8) and also have built into them the correct boundary condition that they vanish on the faces of the box. Imagine we know that we have introduced the particle into the box in such a way that its wave function at $t = 0$ is $\Psi(\boldsymbol{r}, 0) = f(\boldsymbol{r})$. (In practice, it would be very hard to determine $f(\boldsymbol{r})$.) Then

$$f(\boldsymbol{r}) = \left(\frac{8}{abc}\right)^{1/2} \sum_{q,r,s=1}^{\infty} c_{qrs} \sin\frac{q\pi x}{a} \sin\frac{r\pi y}{b} \sin\frac{s\pi z}{c}.$$

This is just a triple Fourier series for $f(\boldsymbol{r})$, which can be inverted by standard methods to give the constants C_{qrs}. If $f(\boldsymbol{r})$ is known, the complete wave function (3.7), for all $t>0$, is thus determined in the form of an infinite series.

The one-dimensional square well

We now consider a one-dimensional problem, where both the potential and the wave function are functions of one of the three coordinates only, say x. We investigate the bound states corresponding to the square-well potential

$$\begin{aligned} V &= 0 \qquad |x|>\tfrac{1}{2}b, \\ &= -U \qquad |x|<\tfrac{1}{2}b, \end{aligned} \tag{3.8}$$

with U positive. This potential represents a very crude one-dimensional model of the potential experienced by an electron in an atom, due to the nucleus. See figure 3.1.

With the potential (3.8), the time-independent Schrödinger equation (2.7) becomes

$$\begin{aligned} (-\hbar^2/2m)(\mathrm{d}^2\psi/\mathrm{d}x^2) &= E\psi \qquad |x|>\tfrac{1}{2}b, \\ (-\hbar^2/2m)(\mathrm{d}^2\psi/\mathrm{d}x^2) - U\psi &= E\psi \qquad |x|<\tfrac{1}{2}b. \end{aligned} \tag{3.9}$$

The solution to the first equation is a linear combination of e^{ikx} and e^{-ikx}, where $E=\hbar^2k^2/2m$. However, for a bound state the particle should be more or less confined to values of x in the vicinity of the potential, and so we should impose the boundary condition $|\psi|\to 0$ as $x\to\pm\infty$. Hence we reject the case $E>0$, since when k is real the solutions do not satisfy this requirement. To satisfy the boundary condition, we must have $E<0$. If we write, with α and $\beta>0$,

$$E=-\hbar^2\alpha^2/2m, \qquad E+U=\hbar^2\beta^2/2m, \tag{3.10}$$

Figure 3.1. (*a*) The one-dimensional square-well potential. (*b*) The Coulomb potential produced by a charged atomic nucleus.

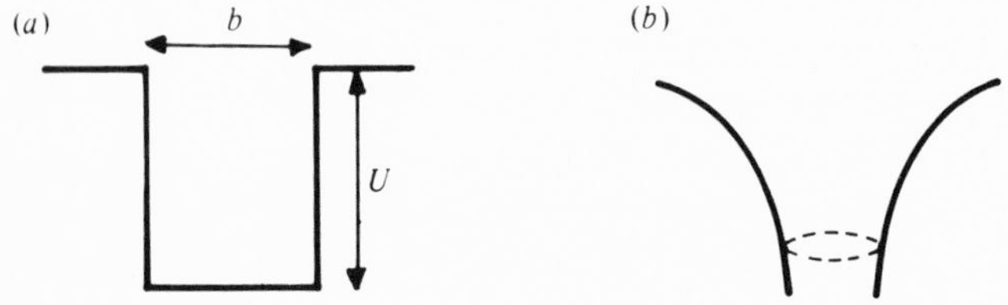

the required solutions to the first equation in (3.9) are

$$\begin{aligned} \psi(x) &= A\,\mathrm{e}^{-\alpha x} \qquad x > \tfrac{1}{2}b, \\ &= B\,\mathrm{e}^{\alpha x} \qquad x < -\tfrac{1}{2}b, \end{aligned} \tag{3.11a}$$

where A and B are constants. (We have rejected a term $\mathrm{e}^{\alpha x}$ for $x > \frac{1}{2}b$, and a term $\mathrm{e}^{-\alpha x}$ for $x < -\frac{1}{2}b$, so as to make $|\psi| \to 0$ as $x \to \pm\infty$). The solution to the second equation in (3.9) is

$$\psi(x) = C \sin \beta x + D \cos \beta x, \tag{3.11b}$$

where C and D are constants.

We now impose the boundary conditions that ψ and $\mathrm{d}\psi/\mathrm{d}x$ are continuous at $x = \pm\frac{1}{2}b$:

$$\begin{aligned} A\,\mathrm{e}^{-\frac{1}{2}\alpha b} &= C \sin \tfrac{1}{2}\beta b + D \cos \tfrac{1}{2}\beta b \\ -\alpha A\,\mathrm{e}^{-\frac{1}{2}\alpha b} &= C\beta \cos \tfrac{1}{2}\beta b - D\beta \sin \tfrac{1}{2}\beta b \\ B\,\mathrm{e}^{-\frac{1}{2}\alpha b} &= -C \sin \tfrac{1}{2}\beta b + D \cos \tfrac{1}{2}\beta b \\ \alpha B\,\mathrm{e}^{-\frac{1}{2}\alpha b} &= C\beta \cos \tfrac{1}{2}\beta b + D\beta \sin \tfrac{1}{2}\beta b. \end{aligned}$$

By eliminating A, B, C, D from these equations, we obtain a relation between β and α. We can manipulate the equations as they stand, but the algebra is somewhat reduced if we first realise that the possible wave functions for the system fall into two classes (see problem 3.6):

(a) Even-parity wave functions, which are symmetric about the origin, $\psi(x) = \psi(-x)$. For these, from (3.11), $A = B$ and $C = 0$, and we find that

$$\alpha = \beta \tan \tfrac{1}{2}\beta b. \tag{3.12a}$$

(b) Odd-parity wave functions, which are antisymmetric about the origin, $\psi(x) = -\psi(-x)$. For these, $A = -B$ and $D = 0$, and

$$\alpha = -\beta \cot \tfrac{1}{2}\beta b. \tag{3.12b}$$

The reason for the existence of these two classes of wave function is that the potential is symmetric about $x = 0$.

If we combine either (3.12a) or (3.12b) with (3.10), we obtain a discrete set of allowed values for the energy E. Each of these lies in the range $0 > E > -U$; that is, the bound states correspond to energies lying within the well. The equations (3.10) and (3.12) can only be solved by numerical methods, but we can get some information about the solutions by graphical methods. If we eliminate α between (3.10) and (3.12), we find that either

$$\tan^2 \tfrac{1}{2}\beta b = -1 + 2mU/(\hbar^2\beta^2) \quad \text{(for even-parity solutions)}$$

or (3.13)

$$\cot^2 \tfrac{1}{2}\beta b = -1 + 2mU/(\hbar^2\beta^2) \quad \text{(for odd-parity solutions).}$$

We have sketched the two sides of these equations, against β, in figure 3.2. In each case the two curves intersect in just a finite number of points, so resulting in a finite set of allowed values for E. The number of such allowed values depends on the magnitudes of U and b. One can show that if $U<0$, there are no bound-state solutions. This is not unexpected: the potential is then repulsive instead of attractive, and so has no bound states.

The linear harmonic oscillator

We now consider another one-dimensional problem, for which the potential is

$$V(x) = \tfrac{1}{2}m\omega^2 x^2. \tag{3.14}$$

In classical dynamics, this would correspond to an attractive force whose magnitude is proportional to the distance from the origin, with simple harmonic solution $x = \sin \omega t$ or $\cos \omega t$. In classical physics most

Figure 3.2. Solution to (3.13).

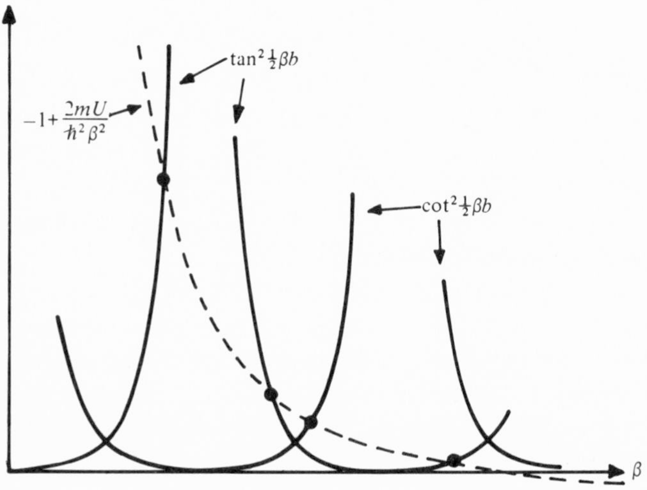

oscillating systems are simple harmonic, at least if the amplitude of oscillation is sufficiently small. For this reason the simple harmonic oscillator plays a central role in classical physics, and the same is true in quantum physics. So the mathematics that follows, with suitable adaptations, applies also, for example, to the quantisation of the electromagnetic field, since according to classical electrodynamics the electromagnetic field in a light wave oscillates harmonically at each point of space. Thus, just as we shall find that in quantum mechanics the energy of a simple linear oscillator with the potential (3.14) can only take certain discrete values, so also the energy in an electromagnetic field appears in discrete quanta. These quanta are what we know as photons.

The time-independent Schrödinger equation for the potential (3.14) is

$$(-\hbar^2/2m)(\mathrm{d}^2\psi/\mathrm{d}x^2)+(\tfrac{1}{2}m\omega^2x^2-E)\psi=0. \qquad (3.15)$$

To this is applied the boundary condition that $\psi\to 0$ as $x\to\pm\infty$, as in the previous problem. This has the consequence that E must take one of the values

$$E_n=(n+\tfrac{1}{2})\hbar\omega, \qquad n=0, 1, 2, \ldots \qquad (3.16)$$

For other values of E the equation has no solutions obeying the boundary condition.

We give an outline proof of this in appendix A. We show there that the corresponding wave functions are

$$\psi_n(x)=\mathrm{e}^{-m\omega x^2/2\hbar}H_n(x), \qquad (3.17)$$

where $H_n(x)$ is a polynomial of degree n in x (it is known as the Hermite polynomial). If n is even, $H_n(x)$ contains only even powers of x, so that $\psi_n(-x)=\psi_n(x)$; if n is odd, $H_n(x)$ contains only odd powers, so that $\psi_n(-x)=-\psi_n(x)$. As in the case of the square well, the potential is symmetric about $x=0$, and so we again find that the solutions have definite parity.

Notice that, according to (3.16), the lowest possible energy of the oscillator is $E_0=\tfrac{1}{2}\hbar\omega$. This is known as *zero-point energy*. In classical mechanics the lowest possible energy is, of course, zero. It corresponds to the particle being at rest at the origin, but in quantum mechanics this is not allowed by the *uncertainty principle*, which we explain in the next chapter.

The time-dependent wave function corresponding to (3.17) is

$$\Psi_n(x, t) = \mathrm{e}^{-m\omega x^2/2\hbar} H_n(x)\,\mathrm{e}^{-\mathrm{i}(n+\frac{1}{2})\omega t}.$$

In general, if the oscillator is somehow set in motion, it will not be in a stationary state. We may write general solutions to the time-dependent Schrödinger equation, satisfying the appropriate boundary conditions, by superposing stationary-state solutions:

$$\Psi(x, t) = \sum_{n=0}^{\infty} c_n \Psi_n(x, t), \tag{3.18}$$

where the constants c_n may take any values. For arbitrary choices of these constants, the probability density at each point x oscillates in time with period $2\pi/\omega$, though the oscillation is not simple harmonic. Notice that although each Ψ_n is a stationary-state wave function and so satisfies the time-independent Schrödinger equation, this is not true of the wave function Ψ that we get by superposing them on each other. This wave function satisfies only the time-dependent Schrödinger equation; it does not correspond to a stationary state.

The tunnel effect

An important phenomenon of quantum mechanics, which is not allowed by the laws of classical physics and which is the basis of some of the applications in the later chapters of this book, is the tunnel effect. Consider, for simplicity, a one-dimensional potential, similar to the square well but opposite in sign:

$$\begin{aligned} V &= 0 \qquad x<0 \quad \text{and} \quad x>a \\ &= U \qquad 0<x<a, \end{aligned} \tag{3.19}$$

with U positive. In classical mechanics a particle approaching the origin from $x<0$ would see this as a barrier which it could surmount only if its energy were greater than U. However, in quantum mechanics a beam of particles will be partially transmitted through this barrier even if the energy of each particle is less than U; that is, some of the particles pass on to $x=+\infty$, while others are reflected back towards $x=-\infty$. In fact, this happens also when $E>U$.

Suppose that the incoming beam of particles has energy E and has the time-independent wave function

$$\psi^{\mathrm{in}} = A\,\mathrm{e}^{\mathrm{i}kx} \tag{3.20}$$

where, as we discussed in chapter 2, the constant A depends on the density of particles in the beam. For $x<0$ the Schrödinger equation is

$$(-\hbar^2/2m)(\mathrm{d}^2\psi/\mathrm{d}x^2)=E\psi, \tag{3.21}$$

so that if ψ^{in} is to satisfy it, $\hbar^2k^2/2m=E$. In the region of the barrier, $0<x<a$, the Schrödinger equation is

$$(-\hbar^2/2m)(\mathrm{d}^2\psi/\mathrm{d}x^2)+U\psi=E\psi, \tag{3.22}$$

so that there the wave function is

$$B\,\mathrm{e}^{\kappa x}+C\,\mathrm{e}^{-\kappa x} \tag{3.23}$$

where $\hbar^2\kappa^2/2m=U-E$, so that if $E<U$, κ is real.

There are two boundary conditions at each edge of the barrier: ψ and $\mathrm{d}\psi/\mathrm{d}x$ are continuous at $x=0$ and at $x=a$. Applying these boundary conditions gives us four equations, but so far we only have two unknowns, B and C, k and κ being determined once we have chosen E. In order to be able to satisfy the boundary conditions we must introduce two more unknowns, so we need a transmitted wave in $x>a$,

$$\psi^{\mathrm{tr}}(x)=T\,\mathrm{e}^{\mathrm{i}kx}, \tag{3.24}$$

and a reflected wave in $x<0$,

$$\psi^{\mathrm{ref}}(x)=R\,\mathrm{e}^{-\mathrm{i}kx},$$

where T and R are constants. The complete wave function is then (3.24) in $x>a$, (3.23) in $0<x<a$ and

$$\psi(x)=A\,\mathrm{e}^{\mathrm{i}kx}+R\,\mathrm{e}^{-\mathrm{i}kx} \tag{3.25}$$

in $x<0$.

Applying the boundary conditions at $x=0$ and $x=a$ now gives us

$$\begin{aligned} A+R&=B+C\\ \mathrm{i}k(A-R)&=\kappa(B-C)\\ T\,\mathrm{e}^{\mathrm{i}ka}&=B\,\mathrm{e}^{\kappa a}+C\,\mathrm{e}^{-\kappa a}\\ \mathrm{i}kT\,\mathrm{e}^{\mathrm{i}ka}&=\kappa(B\,\mathrm{e}^{\kappa a}-C\,\mathrm{e}^{-\kappa a}). \end{aligned} \tag{3.26}$$

We may solve these equations, to determine R, T, B and C in terms of the incoming amplitude A. In particular, we find that

$$T=\frac{4\mathrm{i}\kappa kA}{\mathrm{e}^{\mathrm{i}ka}[\mathrm{e}^{\kappa a}(k+\mathrm{i}\kappa)^2+\mathrm{e}^{-\kappa a}(\kappa+\mathrm{i}k)^2]}, \tag{3.27}$$

so that there is indeed a non-zero transmitted wave. Notice that there is a solution for any positive value of E; this is to be contrasted with bound-state problems, where only certain discrete values of E turn out to be allowed.

If one solves also for R, one finds that

$$|R|^2+|T|^2=|A|^2. \tag{3.28}$$

This equation is the condition that every particle is either reflected or transmitted, and none is lost. To see this, we use the particle flux (2.18); in this one-dimensional problem it is

$$j(x)=\frac{\hbar}{2mi}\left(\psi^*\frac{d\psi}{dx}-\psi\frac{d\psi^*}{dx}\right). \tag{3.29}$$

In a stationary state, where the particle density does not vary with time, this must be independent of x, so that no particles disappear. (This particular simplicity of the consequences of the particle-conservation requirement is peculiar to one-dimensional problems; see problem 2.4.) In $x<0$, (3.25) gives

$$j=(\hbar k/m)(|A|^2-|R|^2), \tag{3.30a}$$

and in $x>a$ (3.24) gives

$$j=(\hbar k/m)|T|^2. \tag{3.30b}$$

Putting (3.30a) and (3.30b) equal to each other, we retrieve (3.28). Notice that, according to (3.24), in $x>a$ the particle density is $|\psi|^2=|T|^2$. The factor $\hbar k/m$ in (3.30b) is the speed of the particles (see the relation between wave number and momentum in (2.2)), so that (3.30b) says that, as we expect, the particle flux is equal to the particle density times the speed. The form (3.30a), applicable in $x<0$, is more complicated because some of the particles are moving in one direction, the rest in the other. In a more general situation, where the particles do not have a single speed, the form of j does not have such a simple interpretation.

The delta-function potential

Our discussion of the tunnel effect, and of bound states in one dimension, was in terms of the simplest possible potential, the square barrier or square well. These form prototypes for repulsive or attractive potentials in one dimension. In many respects the solutions for other, more realistic, potentials in one, or indeed in three, dimensions have similar properties. However, the solutions for even these simple potentials involve a fair amount of algebra, so it is often simpler to consider a mathematical limit that is easier to work with.

Consider the square-barrier potential (3.19) and let $a \to 0$, with at the same time $U \to \infty$, in such a way that aU remains fixed and equal to λ, say. The resulting potential is just

$$V(x) = \lambda\delta(x). \tag{3.31}$$

(For a description of the delta function, see appendix B.) To find the stationary states, we have to solve the time-independent Schrödinger equation

$$-\frac{\hbar^2}{2m}\frac{\mathrm{d}^2\psi}{\mathrm{d}x^2} + \lambda\delta(x)\psi(x) = E\psi(x). \tag{3.32}$$

This equation is to be valid for all x. From it, we conclude that $\psi(x)$ is continuous even at $x = 0$. For suppose that $\psi(x)$ were discontinuous at $x = 0$. Then $\psi'(x)$ would contain a delta-function part (or a worse singularity if the discontinuity of ψ is infinite) and so ψ'' would contain the derivative of $\delta(x)$ (or worse). This cannot be, because such a singularity in ψ'' is not balanced by a similar singularity in any other term of the Schrödinger equation (3.32). However, $\psi'(x)$ must be discontinuous at $x = 0$, so that $\psi''(x)$ contains a delta-function singularity that balances that of the second term of the Schrödinger equation.

Take the wave function in $x < 0$ to be an incoming plane wave together with a reflected wave, just as in (3.25):

$$\psi(x) = A\,\mathrm{e}^{ikx} + R\,\mathrm{e}^{-ikx}. \tag{3.33a}$$

Let the transmitted wave be as in (3.24):

$$\psi(x) = T\,\mathrm{e}^{ikx} \qquad x > 0. \tag{3.33b}$$

In order that the wave function be continuous at $x = 0$,

$$A + R = T. \tag{3.34}$$

To obtain the correct discontinuity of $\psi'(x)$ at $x = 0$, integrate the Schrödinger equation (3.32) from $x = -\varepsilon$ to $x = +\varepsilon$ and then let $\varepsilon \to 0$. The various terms of the Schrödinger equation give

$$\frac{-\hbar^2}{2m}\int_{-\varepsilon}^{\varepsilon} \mathrm{d}x\,\psi''(x) \to \frac{-\hbar^2}{2m}\,\mathrm{disc}\,\psi'$$

$$\lambda\int_{-\varepsilon}^{\varepsilon} \mathrm{d}x\,\delta(x)\psi(x) = \lambda\psi(0)$$

$$E\int_{-\varepsilon}^{\varepsilon} \mathrm{d}x\,\psi(x) \sim 2\varepsilon E\psi(0) \to 0,$$

where disc ψ' denotes the discontinuity of ψ' at $x=0$, that is, the value of $\psi'(x)$ at x just greater than zero minus its value at x just less than zero. Hence

$$(-\hbar^2/2m)\text{ disc }\psi' + \lambda\psi(0) = 0 \tag{3.35}$$

and, from the wave function (3.33),

$$(-\hbar^2/2m)[\mathrm{i}kT - \mathrm{i}k(A-R)] + \lambda T = 0.$$

With (3.34), we thus find that

$$\begin{aligned} T &= m\lambda A/(\mathrm{i}\hbar^2 k - m\lambda) \\ R &= \mathrm{i}\hbar^2 kA/(\mathrm{i}\hbar^2 k - m\lambda). \end{aligned} \tag{3.36}$$

It is straightforward to check that the particle-conservation condition (3.28) is satisfied.

Problems

3.1 Sketch the wave functions for the one-dimensional square-well potential (3.8).

It is found that the average distance of the particle from the centre of the well can be rather greater than the width of the well. How is this possible, in terms of your sketches?

3.2 A one-dimensional harmonic oscillator carries charge e and is placed in a uniform electric field $\mathscr{E}$, so that the potential becomes

$$V(x) = \tfrac{1}{2}m\omega^2 x^2 - e\mathscr{E}x.$$

Show that each energy level is reduced by $e^2\mathscr{E}^2/2m\omega^2$. What are the new wave functions? [*Hint*: replace x by a new variable.]

3.3 By consulting appendix A, calculate the $n=0$ and $n=2$ wave functions of the linear harmonic oscillator (leave them un-normalised).

Verify that they are orthogonal, that is,

$$\int_{-\infty}^{\infty} \mathrm{d}x\, \psi_0(x)\psi_2(x) = 0.$$

3.4 A beam of particles is incident on (i) a very deep square-well potential, (ii) a narrow barrier ($ka \ll 1$) whose height is equal to twice the kinetic energy of each particle. Show that in case (i) the particles are nearly all reflected, while in case (ii) they are nearly all transmitted.

Notice how different these results are from those which obtain in the corresponding situations in classical mechanics.

3.5 Find the energy level of the bound state for the potential $V(x) = -\mu\delta(x)$. Verify that this agrees with that for the square-well potential (3.8) in the limit $U\to\infty$, with Ub fixed and equal to μ.

3.6 A potential $V(x)$ is such that $V(x)=V(-x)$ and each bound-state energy level corresponds to only one independent wave function. Show that each of these wave functions has either even parity or odd parity. [*Hint*: the first step in the proof is to show that if $\psi(x)$ is a solution of the Schrödinger equation, so also is $\psi(-x)$, with the same value of E].

4

The superposition principle

Linear operators

In the examples of chapter 3, we took linear combinations of solutions to the Schrödinger equation and so constructed further solutions: see (3.7) and (3.18). The reason that this works is that $i\hbar\partial/\partial t$ and the Hamiltonian operator H, which appear in the Schrödinger equation, are linear, and, furthermore, the boundary and continuity conditions on the wave function Ψ are linear.

By linear boundary or continuity conditions, we mean that if Ψ_1 and Ψ_2 are two functions that satisfy them, then so does the function $(c_1\Psi_1+c_2\Psi_2)$, where c_1 and c_2 are any complex numbers. This is evidently the case for wave functions, which we normally require to vanish outside some region of space, or at infinity.

By a linear operator $\hat{Q}$, we mean that

$$\hat{Q}(c_1\Psi_1+c_2\Psi_2)=c_1\hat{Q}\Psi_1+c_2\hat{Q}\Psi_2. \tag{4.1}$$

An example of an operator that is not linear is the operator $\hat{S}$ that squares the wave function:

$$\hat{S}\Psi=\Psi^2.$$

In chapter 2 we said that one of the basic assumptions of quantum mechanics is that observables correspond to operators. This assumption is usually supplemented by the further requirement that the operators be linear, and in this chapter we explore some of the consequences of this. Notice that the operators we have met so far, $-i\hbar\nabla$ corresponding to momentum and H corresponding to energy, are linear.

The consequence of linearity is that if we superpose wave functions that describe possible states of a system, we get another wave function that describes a possible state. It is often useful, as in our examples of (3.7) and (3.18), to superpose stationary-state solutions. A basic mathematical property that can be proved is that these stationary-state wave functions are *complete*. This means that any allowable $\Psi(\boldsymbol{r}, t)$, that is, any wave function that satisfies the time-dependent Schrödinger equation and the appropriate boundary conditions for the system, can be written as a sum of stationary-state wave functions

$$\Psi(\boldsymbol{r}, t) = \sum_n c_n \psi_n(\boldsymbol{r})\, \mathrm{e}^{-\mathrm{i}E_n t/\hbar}. \tag{4.2}$$

Wave packets

As an example, consider a free particle moving in the x-direction. The stationary states are the plane waves

$$N(p)\,\mathrm{e}^{\mathrm{i}px/\hbar - \mathrm{i}p^2t/2m\hbar}, \tag{4.3}$$

corresponding to a definite value p of the momentum and a definite value $p^2/2m$ of the energy. $N(p)$ is a constant. As the squared modulus of this wave function is independent of x, there is equal probability of finding the particle at each point x. To construct a wave function corresponding to a particle that is localised in space, we must take a superposition of the stationary states (4.3). These are labelled by the momentum p, which is here a continuous variable. This is because, as we explained in chapter 2, the plane wave (4.3) describes a mathematical limit where the volume Ω of the system has been allowed to become infinite. For finite Ω the superposition (4.2) would be a sum over discrete values of p (see (3.7)). As $\Omega \to \infty$ this sum becomes an integral:

$$\Psi(x, t) = \int \mathrm{d}p\, N(p)\,\mathrm{e}^{\mathrm{i}px/\hbar - \mathrm{i}p^2t/2m\hbar}. \tag{4.4}$$

Here N may take a different value for each value of p: it is now a function of p. The resulting wave function is not a stationary state; if we wish to localise the particle in space, it does not have definite momentum or energy. Rather, (4.4) expresses the wave function as a superposition of components having different values of momentum p and energy $p^2/2m$.

Setting $t = 0$, we have

$$\Psi(x, 0) = \int \mathrm{d}p\, N(p)\,\mathrm{e}^{\mathrm{i}px/\hbar}. \tag{4.5}$$

That is, $N(p)$ is the Fourier transform of $\Psi(x, 0)$ (see appendix B). By inverting the transform, we may express $N(p)$ in terms of $\Psi(x, 0)$:

$$N(p)=\frac{1}{2\pi\hbar}\int dx\ \Psi(x, 0)\ e^{-ipx/\hbar}. \tag{4.6}$$

Hence if we know the wave function Ψ at $t=0$, we can calculate $N(p)$. Inserting the result into (4.4), we can then find the wave function Ψ at other times t.

As a simple example, suppose that we take a *Gaussian*:

$$\Psi(x, 0)=N_0\ e^{-(x-x_0)^2/a^2+ip_0x/\hbar}. \tag{4.7}$$

Here N_0 is a normalisation constant, chosen so that $\int dx|\Psi(x, 0)|^2=1$, which requires that

$$N_0=(\tfrac{1}{2}a^2\pi)^{-1/4}.$$

The wave function (4.7) is called a *wave packet*: the probability density $|\Psi(x, 0)|^2$ is localised around the point $x=x_0$. If we prepare the state a large number of times and each time measure the position of the particle at $t=0$, because $|\Psi|^2$ is the probability density the average value will be

$$\langle x\rangle=\int dx\, x|\Psi(x, 0)|^2=x_0. \tag{4.8}$$

The results of the separate measurements will be scattered around this average value. A measure of the width of their distribution is the quantity Δx, defined by

$$(\Delta x)^2=\int dx(x-\langle x\rangle)^2\,|\Psi(x, 0)|^2. \tag{4.9}$$

Δx is called the width of the wave packet. With (4.8) we find that, at $t=0$, $\Delta x=\frac{1}{2}a$.

If we insert (4.7) into (4.6), we obtain an integral whose evaluation is discussed in appendix B. It gives

$$N(p)=(8\hbar^4\pi^3/a^2)^{-1/4}\ e^{-(p-p_0)^2a^2/4\hbar^2-ipx_0/\hbar}. \tag{4.10}$$

This has the same functional form as $\Psi(x, 0)$, a property that is peculiar to the Gaussian. One can check that $\int dp\ |\phi(p)|^2=1$, where

$$\phi(p)=[1/(2\pi\hbar)^{1/2}]N(p).$$

The interpretation is that, just as $|\Psi|^2$ is the probability density in position space, so $|\phi|^2$ is the probability density in momentum space.

That is, if we measure the momentum, the probability of obtaining a result in the interval $(p, p+dp)$ is just $|\phi(p)|^2\,dp$. If we prepare the state many times and measure the momentum each time, the average value so obtained will be

$$\langle p\rangle = \int \mathrm{d}p\, p|\phi(p)|^2 = p_0, \tag{4.11}$$

and the distribution of values obtained will have a width Δp defined by

$$(\Delta p)^2 = \int \mathrm{d}p (p-\langle p\rangle)^2\, |\phi(p)|^2. \tag{4.12}$$

With (4.7), we find that $\Delta p = \hbar/a$. Hence at $t=0$

$$\Delta x\ \Delta p = \tfrac{1}{2}\hbar.$$

There is a general principle, the *Heisenberg uncertainty principle*, which says that for any wave function, Δx and Δp, defined by (4.9) and (4.12), satisfy

$$\Delta x\ \Delta p \geqslant \tfrac{1}{2}\hbar. \tag{4.13}$$

Δx and Δp are the uncertainties in the position and momentum associated with the wave packet. This means that we cannot know the position and momentum together to better than to a certain degree of accuracy. The allowed degree of accuracy depends on the shape of the wave packet; in this sense a Gaussian is the best possible since it gives $\Delta x\ \Delta p = \frac{1}{2}\hbar$ rather than greater than $\frac{1}{2}\hbar$. Notice that in the case of a plane wave, the momentum takes a definite value, so $\Delta p = 0$. However, since $|\Psi|^2$ has the same value for all x, in this case the integral (4.9) diverges and $\Delta x = \infty$.

Ehrenfest's theorem

Having obtained $N(p)$ in (4.10), we may now use (4.4) to calculate $\Psi(x, t)$ for $t \neq 0$ (see problem 4.2). The result is that $\Psi(x, t)$ remains Gaussian in shape at subsequent times t. The average position $\langle x\rangle$ of the particle moves with velocity p_0/m, and the width Δx of the wave packet increases with increasing time. The reason that the wave packet spreads out as t increases is that the different momentum components of the wave function correspond to different velocities p/m.

There is a theorem, due to Ehrenfest, that states that whatever the shape of the wave packet, and for a general potential V,

$$(\mathrm{d}/\mathrm{d}t)\langle \boldsymbol{r}\rangle = \langle \boldsymbol{p}\rangle/m \tag{4.14a}$$

and

$$(\mathrm{d}/\mathrm{d}t)\langle \boldsymbol{p}\rangle = \langle -\boldsymbol{\nabla} V\rangle, \tag{4.14b}$$

where

$$\langle \boldsymbol{\nabla} V \rangle = \int d^3\boldsymbol{r} \boldsymbol{\nabla} V \, |\Psi|^2$$

(see problem 4.3). For a sufficiently narrow wave packet, $\langle \boldsymbol{\nabla} V \rangle$ is almost equal to the value of $\boldsymbol{\nabla} V$ at $\boldsymbol{r} = \langle \boldsymbol{r} \rangle$. Then the two equations (4.14) show that the wave packet as a whole moves like a classical particle. This does not mean that it describes a classical situation because, as we have seen, the momentum and position of the particle cannot be simultaneously known with absolute certainty. Crudely speaking, the quantum-mechanical description is very close to the classical description if the widths Δr, Δp of the wave packet in position and momentum are small. This statement is not precise because we have not specified a length L and a momentum P with which we must compare Δr and Δp in order to decide whether they are indeed small. L and P will be parameters somehow typical of the system under investigation. Because we know from the uncertainty principle that $\Delta r \, \Delta p \geqslant \frac{1}{2}\hbar$, it is necessary that

$$LP \gg \hbar$$

if a classical description is to be a good approximation.

Hermitian operators

We have said that it is assumed in quantum mechanics that to each observable Q there corresponds a linear operator $\hat{Q}$. We now introduce an additional assumption about this operator: that it is Hermitian.

Let Ψ_1 and Ψ_2 be any two of the allowed wave functions for the system under discussion. We write

$$\int d^3\boldsymbol{r} \, \Psi_1^*(\hat{Q}\Psi_2) = (\Psi_1, \hat{Q}\Psi_2). \tag{4.15}$$

In this definition the integration is, as usual, over all space. The operator $\hat{Q}$ is said to be *Hermitian* if, for all choices of Ψ_1 and Ψ_2,

$$(\Psi_1, \hat{Q}\Psi_2) = (\Psi_2, \hat{Q}\Psi_1)^*, \tag{4.16a}$$

that is, if

$$\int d^3\boldsymbol{r} \, \Psi_1^*(\hat{Q}\Psi_2) = \int d^3\boldsymbol{r} (\hat{Q}\Psi_1)^*\Psi_2. \tag{4.16b}$$

As an example, the Hamiltonian operator H is Hermitian. To show this, we use the fact that $V = V^*$, since the potential V is real, so that

$$(\Psi_1, H\Psi_2) - (\Psi_2, H\Psi_1)^*$$
$$= \int d^3\boldsymbol{r} \left[\Psi_1^*\left(-\frac{\hbar^2}{2m}\nabla^2\Psi_2\right) - \left(-\frac{\hbar^2}{2m}\nabla^2\Psi_1^*\right)\Psi_2\right].$$

Just as in (2.15), we may transform this to an integral over the surface enclosing the volume of integration and conclude that for a closed system, where the wave functions vanish outside some volume Ω_0, the integral vanishes. As we explained in chapter 2, an unbounded system, corresponding to the mathematical limit $\Omega_0 \to \infty$, is a little more subtle, and we shall not discuss it.

An important property of a Hermitian operator is that its eigenvalues are real, as we now show. For if $\hat{Q}\Psi = q\Psi$, where q is one of the eigenvalues, then

$$(\Psi, \hat{Q}\Psi) = q(\Psi, \Psi), \tag{4.17}$$

where

$$(\Psi_1, \Psi_2) = \int d^3\boldsymbol{r}\, \Psi_1^*\Psi_2.$$

Take the complex conjugate of (4.17) and use (4.16*a*) for the case $\Psi_1 = \Psi_2 = \Psi$:

$$(\Psi, \hat{Q}\Psi) = q^*(\Psi, \Psi).$$

Hence, since $(\Psi, \Psi) > 0$, $q = q^*$. Because, as we explained in chapter 2, the eigenvalues of the operator $\hat{Q}$ corresponding to an observable Q are possible values obtained in measurements of Q, it is natural to impose the restriction that $\hat{Q}$ be Hermitian so as to ensure that these values are real. We shall argue below that whenever we measure Q, the result is one of the eigenvalues of $\hat{Q}$.

Another important property of a Hermitian operator is that two eigenfunctions Ψ_1 and Ψ_2, corresponding to different eigenvalues q_1 and q_2, are *orthogonal*. By this we mean that

$$(\Psi_1, \Psi_2) = 0.$$

To show this, since by definition $\hat{Q}\Psi_1 = q_1\Psi_1$ and $\hat{Q}\Psi_2 = q_2\Psi_2$, we write the chain of equations

$$q_2(\Psi_1, \Psi_2) = (\Psi_1, \hat{Q}\Psi_2) = (\Psi_2, \hat{Q}\Psi_1)^* = q_1^*(\Psi_2, \Psi_1)^* = q_1(\Psi_1, \Psi_2).$$

Here we have used (4.16*a*) and the fact that q_1 is real. So if $q_1 \neq q_2$, we have the required result. If $q_1 = q_2$, we cannot conclude that Ψ_1 and Ψ_2

are orthogonal. But in this case, any linear combination $(c_1\Psi_1 + c_2\Psi_2)$ is also an eigenfunction of $\hat{Q}$, with the same eigenvalue, and we can choose two linear combinations that are orthogonal. If we normalise each eigenfunction, so that it satisfies $(\Psi, \Psi) = 1$, we then have an *orthonormal set*:

$$(\Psi_r, \Psi_s) = \delta_{rs}, \tag{4.18}$$

where $\delta_{rs} = 1$ if $r = s$ and $\delta_{rs} = 0$ if $r \neq s$.

A basic *mathematical* assumption, which is not altogether straightforward to check in practice, is that the set is *complete*. By this we mean that any allowable wave function Ψ of the system can be expanded as a linear combination of the eigenfunctions Ψ_r of $\hat{Q}$:

$$\Psi = \sum_r c_r \Psi_r. \tag{4.19}$$

If we want Ψ to be correctly normalised, so that $(\Psi, \Psi) = 1$, the coefficients c_r will satisfy

$$\sum_r |c_r|^2 = 1. \tag{4.20}$$

To verify this, use the orthonormality condition (4.18). From this condition we find also that

$$(\Psi_s, \Psi) = c_s. \tag{4.21}$$

Operators and observables

We explained in chapter 2 that it is assumed that when the value of an observable Q is known to be q, the system is in a state whose wave function is an eigenfunction of the corresponding operator $\hat{Q}$, with eigenvalue q. We say that the system is then in an *eigenstate* of $\hat{Q}$. This assumption implies that any measurement of Q will leave the system in an eigenstate of $\hat{Q}$. This follows if we assume that quantum physics is not so capricious that, if with suitable care we make the same measurement of Q twice, in rapid succession, we expect to get different answers. The two measurements must be close enough in time so that we can be sure that no outside influence has disturbed the system, and that the natural time development of the system has not resulted in a change in the value of Q. After the first measurement, we know what will be the result of the second measurement and so, according to the assumption, the first measurement must have left the system in an eigenstate of $\hat{Q}$ and the result of the measurement is the corresponding eigenvalue q.

In general, the system will not be in an eigenstate of $\hat{Q}$ before the first measurement. So the first measurement must have *disturbed* the system, so as to throw it into an eigenstate of $\hat{Q}$.

Suppose that, before any measurement of Q is made, the system is in a state whose wave function is Ψ. The result of measuring Q is one of the eigenvalues of $\hat{Q}$ but, unless Ψ happens to be an eigenstate of $\hat{Q}$, we cannot predict which eigenvalue it will be, that is, which eigenstate the system will jump into because of our act of measurement. However, we can calculate the *probability* of obtaining a particular eigenvalue. One would guess that this probability is related to the coefficients c_r in the expansion (4.19) of Ψ in terms of the eigenfunctions Ψ_r. The most natural assumption is that the probability P_r of throwing the system into the eigenstate Ψ_r by making the measurement, and so obtaining the eigenvalue q_r as the value of Q, is

$$P_r = |c_r|^2 = |(\Psi_r, \Psi)|^2. \tag{4.22}$$

Then, because of the normalisation condition (4.20),

$$\sum_r P_r = 1. \tag{4.23}$$

That is, the probability of getting *some* result for the measurement is 1.

Let us now recapitulate our basic assumptions about observables:

(*a*) To each observable Q there corresponds a linear, Hermitian operator $\hat{Q}$.
(*b*) If the value of Q is known to be q, the system is in an eigenstate of $\hat{Q}$, with eigenvalue q.
(*c*) The result of measuring Q can be any of the eigenvalues of $\hat{Q}$. When the system is in the state Ψ, the probability that a measurement of Q will give the eigenvalue q_r is $P_r = |(\Psi_r, \Psi)|^2$.

If we somehow prepare the same state Ψ many times, and on each occasion measure Q at the same relative instant in time, the average answer will be

$$\sum_r q_r P_r = \sum_r q_r \, |c_r|^2.$$

If we use the definition (4.19) of the c_r, and the orthonormality condition (4.18), we find that

$$\sum q_r P_r = (\Psi, \hat{Q}\Psi). \tag{4.24}$$

This is called the *expectation value* of Q in the state Ψ. It is often written $\langle Q \rangle$. Earlier in this chapter we have already encountered two particular examples, the expectation values of position and momentum:

$$\langle \boldsymbol{r} \rangle = \int \mathrm{d}^3\boldsymbol{r}\ \Psi^*(\hat{\boldsymbol{r}}\Psi) = \int \mathrm{d}^3\boldsymbol{r}\ \boldsymbol{r}|\Psi|^2$$

$$\langle \boldsymbol{p} \rangle = \int \mathrm{d}^3\boldsymbol{r}\ \Psi^*(\hat{\boldsymbol{p}}\Psi) = \int \mathrm{d}^3\boldsymbol{r}\ \Psi^*(-\mathrm{i}\hbar\boldsymbol{\nabla}\Psi). \tag{4.25}$$

The operator $\hat{\boldsymbol{r}}$, corresponding to position, simply multiplies any wave function on which it acts by $\boldsymbol{r}$. It may be verified that the expression (4.25) for $\langle \boldsymbol{p} \rangle$ is equivalent to the definition (4.11) in terms of $\phi(\boldsymbol{p})$.

Commutators

Suppose that we measure an observable Q and then another observable R. The first measurement will throw the system into an eigenstate of $\hat{Q}$. In general, this will not be an eigenstate of $\hat{R}$, so that the second measurement will disturb the system again. Thus, by measuring R, we lose the information we have about Q.

This is another manifestation of the uncertainty principle, which we have already encountered for the particular case of the observables $\boldsymbol{r}$ and $\boldsymbol{p}$. In general, one cannot know the exact values of two observables Q and R simultaneously, because generally their eigenstates are not the same.

However, some pairs of observables may simultaneously be measured exactly. It can be shown that two operators $\hat{Q}$ and $\hat{R}$ have a complete set of common eigenstates if, and only if, their *commutator*,

$$[\hat{Q}, \hat{R}] = \hat{Q}\hat{R} - \hat{R}\hat{Q}, \tag{4.26}$$

vanishes. That is, if we operate on any wave function first with $\hat{Q}$ and then with $\hat{R}$, the result is the same as when we apply the operators in the reverse order. We then say that the operators $\hat{Q}$ and $\hat{R}$ *commute*. So the condition that we may simultaneously know the values of two observables is that the corresponding operators commute.

The generalisation of the uncertainty principle (4.13) for $\Delta x\ \Delta p_x$ to an arbitrary pair of observables can be shown to be (see problem 4.11)

$$\Delta Q\ \Delta R \geqslant \tfrac{1}{2}|(\Psi, [\hat{Q}, \hat{R}]\Psi)|. \tag{4.27}$$

On the right of this inequality there appears the modulus of the expectation value of the commutator of the corresponding operators, in the state for which the measurement is being made. The commutator is, of course, itself an operator. For the particular case of components of position and momentum, it turns out that the operator is a simple number:

$$[\hat{x}_i, \hat{p}_j] = \mathrm{i}\hbar\ \delta_{ij} \tag{4.28}$$

where i and j label the different components, and $\delta_{ij} = 1$ if $i = j$ and 0 if $i \neq j$. To check this, introduce an arbitrary wave function Ψ and let the commutator operate on it:

$$[\hat{x}_i, \hat{p}_j]\Psi = x_i(-\mathrm{i}\hbar\ \partial/\partial x_j)\Psi - (-\mathrm{i}\hbar\ \partial/\partial x_j)(x_i\Psi) = \mathrm{i}\hbar\ \delta_{ij}\Psi.$$

Thus the two sides of the relation (4.28) are equal when applied to any wave function Ψ, which is the same thing as saying that (4.28) is a true identity. Because the commutator (4.28) vanishes when $i \neq j$, we can measure simultaneously the exact values of a given component of $\boldsymbol{r}$ and a different component of $\boldsymbol{p}$. When we take the same components of $\boldsymbol{r}$ and $\boldsymbol{p}$, the commutator (4.28) is $\mathrm{i}\hbar$, and so the general uncertainty relation (4.27) then agrees with (4.13): we cannot measure the exact values together.

The theory of measurement in quantum mechanics is a deep, and even controversial, subject; our discussion has necessarily been brief.

Problems

4.1 Calculate $\langle p\rangle$ and Δp for the Gaussian wave packet (4.7) at $t = 0$ by taking expectation values of the appropriate operators.

4.2 A Gaussian wave packet is given at $t = 0$ by (4.7). Calculate $\Psi(x, t)$ from (4.10) and (4.4) far enough to verify that

$$\langle x\rangle = x_0 + p_0 t/m,$$
$$(\Delta x)^2 = \tfrac{1}{4}a^2 + \hbar^2 t^2/m^2 a^2.$$

4.3 Prove Ehrenfest's theorem, equations (4.14), for a one-dimensional situation where the wave function depends on x and t only.

4.4 Show, from the uncertainty principle, that the minimum energy of a one-dimensional harmonic oscillator must be greater than, or equal to, $\tfrac{1}{2}\hbar\omega$. [*Hint*: write down the expectation value of H and note that both $\langle x\rangle$ and $\langle p\rangle$ are zero because of the symmetry about the origin.]

4.5 A dart of mass 1 kg is dropped from a height of 1 m, the intention being to hit a given point on the ground below. The uncertainty principle imposes a limitation on the accuracy that can be achieved; show that this is only of the order of a tenth of a nuclear diameter. [Neglect the uncertainties in position and momentum in the vertical direction; these produce second-order effects on the accuracy.]

4.6 Show that the momentum operator $-i\hbar\nabla$ is Hermitian for a bounded system.

4.7 Use the divergence theorem (page 15) to show that in a bounded system the expectation value of the one-particle kinetic-energy operator is positive.

By considering the expectation value of the Hamiltonian, deduce that for a particle moving in a potential well of arbitrary shape the lowest bound state has an energy level that is always higher than the bottom of the well.

4.8 $\hat{Q}$ is a Hermitian operator and the wave function Ψ is such that $\hat{Q}\Psi = q\Psi$, where q is a number. Φ is any other wave function. Show that

$$(\Psi, \hat{Q}\Phi) = q(\Psi, \Phi).$$

4.9 A quantum system has a complete orthonormal set of energy eigenstates ψ_n, with different eigenvalues E_n. The operator $\hat{Q}$ corresponds to an observable and is such that

$$\hat{Q}\psi_1 = \psi_2, \qquad \hat{Q}\psi_2 = \psi_1, \qquad \hat{Q}\psi_n = 0 \quad \text{for } n \geqslant 3.$$

Find a complete orthonormal set of eigenfunctions of $\hat{Q}$.

The observable is measured and is found to have the value +1. The system is left undisturbed and after a time t the observable is measured again. Calculate the probability that the value +1 is found again.

4.10 The operator $\hat{Q}$ does not depend explicitly on t. Show that for any solution Ψ of the time-dependent Schrödinger equation

$$i\hbar\frac{d}{dt}(\Psi, \hat{Q}\Psi) = (\Psi, [\hat{Q}, H]\Psi).$$

4.11 We define the *Hermitian conjugate* $\hat{F}^\dagger$ of an operator $\hat{F}$ such that, for any pair of allowed wave functions of the quantum system under discussion, $(\Psi_1, \hat{F}^\dagger\Psi_2) = (\Psi_2, \hat{F}\Psi_1)^*$. Show that

(i) if $\hat{F}$ is hermitian, then $\hat{F}^\dagger = \hat{F}$
(ii) $(\hat{F} + i\hat{G})^\dagger = \hat{F}^\dagger - i\hat{G}^\dagger$
(iii) $(\hat{F}\hat{G})^\dagger = \hat{G}^\dagger\hat{F}^\dagger$
(iv) if $\Phi = \hat{F}\Psi$ then $(\Phi, \Psi) = (\Psi, \hat{F}^\dagger\Psi)$.

The operators $\hat{Q}$ and $\hat{R}$ correspond to observables and λ is any real number. Show that the expectation value of $(\hat{Q} + i\lambda\hat{R})(\hat{Q} - i\lambda\hat{R})$ in any state Ψ is real and $\geqslant 0$. Hence prove that

$$(\Psi, \hat{Q}^2\Psi)(\Psi, \hat{R}^2\Psi) \geqslant \tfrac{1}{4}(\Psi, i[\hat{Q}, \hat{R}]\Psi)^2$$

and so deduce the general uncertainty principle (4.27).

5

The hydrogen atom

Good quantum numbers

In general, if we measure an observable Q at two different times, we shall obtain different eigenvalues of $\hat{Q}$. This is because the system will have undergone a natural time development, governed by the time-dependent Schrödinger equation. (In addition, the system may have been disturbed from outside, such as by a measurement of some other observable.)

The only states whose natural time development is trivial are the stationary states (2.9), for which the particle density $|\Psi|^2$ is constant in time at each point $\boldsymbol{r}$. The stationary states are the eigenstates of the Hamiltonian operator H. If we measure the observable Q, we throw the system into an eigenstate of the corresponding operator $\hat{Q}$. If this state is also an eigenstate of H, its time development is trivial and a subsequent measurement of Q will give the same result as the first measurement. According to the discussion given at the end of the last chapter, in order to ensure that the eigenstates of $\hat{Q}$ and H are the same, we must have the condition

$$[\hat{Q}, H] = 0. \tag{5.1}$$

In these circumstances we say that the observable Q is a good quantum number. Its value can be used to label a stationary state, because it does not change with time. (See also problem 4.9.)

Orbital angular momentum

An observable that is often useful for labelling stationary states is the orbital angular momentum. In classical mechanics the orbital angular momentum of a particle about the origin of coordinates is $\boldsymbol{r} \wedge \boldsymbol{p}$. In quantum mechanics the corresponding operator is usually called $\hbar\boldsymbol{L}$, and it is given by

$$\hbar\boldsymbol{L}=\boldsymbol{r}\wedge\hat{\boldsymbol{p}}=\boldsymbol{r}\wedge(-i\hbar\boldsymbol{\nabla}). \tag{5.2}$$

An important property of the operator $\boldsymbol{L}$ is that its square is closely related to the operator ∇^2 which appears in the Hamiltonian. The calculation of $\boldsymbol{L}^2$ is described in appendix C and the result is that

$$\boldsymbol{L}^2=-\frac{1}{\sin\theta}\frac{\partial}{\partial\theta}\sin\theta\frac{\partial}{\partial\theta}-\frac{1}{\sin^2\theta}\frac{\partial^2}{\partial\phi^2}, \tag{5.3}$$

where θ and ϕ are polar and azimuthal angles of spherical polar coordinates (figure 5.1). In spherical polar coordinates,†

$$\nabla^2=\frac{1}{r}\frac{\partial^2}{\partial r^2}r-\frac{1}{r^2}\boldsymbol{L}^2, \tag{5.4}$$

so that the part of ∇^2 that involves the angles θ and ϕ is just $-\boldsymbol{L}^2/r^2$.

If we now define the polar axis $\theta=0$ of the spherical polar coordinate system to lie along the z-direction, it can be shown from the definition (5.2) of $\boldsymbol{L}$ that its z-component takes the simple form

$$L_z=-i\,\partial/\partial\phi. \tag{5.5}$$

According to (5.3), $\boldsymbol{L}^2$ does not involve ϕ explicitly, only $\partial/\partial\phi$. This means that $\boldsymbol{L}^2$ commutes with L_z, that is, $[\boldsymbol{L}^2, L_z]=0$, so that the two operators have simultaneous eigenfunctions. It is usual to denote these

Figure 5.1. Spherical polar coordinates.

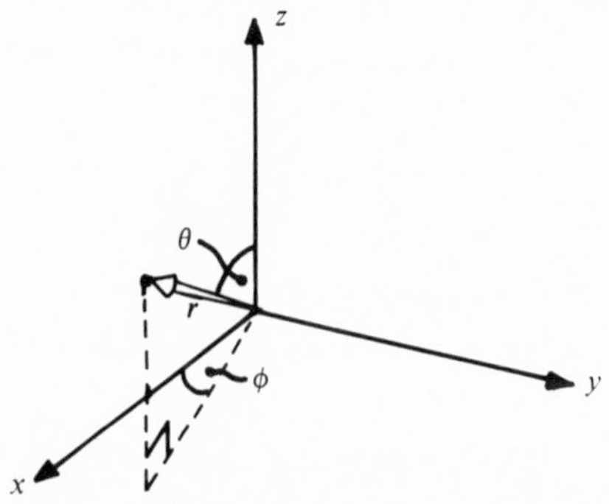

† For the calculation of ∇^2 in spherical polar coordinates, see any textbook on vector calculus, for example H. Jeffreys and B. S. Jeffreys, *Methods of Mathematical Physics*, 3rd edn (Cambridge University Press, 1956).

by $Y_{lm}(\theta, \phi)$, where the labels l and m are defined in terms of the eigenvalues of $\boldsymbol{L}^2$ and $\boldsymbol{L}_z$ as follows:

$$\begin{aligned} L_z Y_{lm} &= m Y_{lm} \\ \boldsymbol{L}^2 Y_{lm} &= l(l+1) Y_{lm}. \end{aligned} \tag{5.6}$$

That is, m is the eigenvalue of L_z but, for a reason that will become apparent shortly (see (5.9) below), the eigenvalue of $\boldsymbol{L}^2$ is written as $l(l+1)$.

The first equation in (5.6), together with (5.5), tells us that Y_{lm} varies with ϕ simply as $e^{im\phi}$. That is,

$$Y_{lm}(\theta, \phi) = P_l^m(\theta)\, e^{im\phi}, \tag{5.7}$$

where $P_l^m(\theta)$ is some function of θ only. Using the second equation of (5.6), together with (5.3), we find that

$$\frac{1}{\sin\theta}\frac{d}{d\theta}\left(\sin\theta\frac{dP_l^m}{d\theta}\right) - \frac{m^2}{\sin^2\theta}P_l^m + l(l+1)P_l^m = 0. \tag{5.8}$$

The solutions of this equation, which is known as Legendre's equation, have been well studied. For most values of l and m the solutions are infinite for $\theta = 0$ or π, and so they do not then yield a physically acceptable wave function. It is found that in order to avoid infinities, and also to ensure that any linear combination of the wave functions Y_{lm} gives a probability density that is single valued (the probability density should not change when ϕ is increased by 2π, since values of ϕ differing by 2π correspond to the same point $\boldsymbol{r}$), we must choose l to be a non-negative integer and restrict m to integers from $-l$ to l. That is, we have the eigenvalues

$$\begin{aligned} l(l+1) \text{ for } \boldsymbol{L}^2, &\quad \text{where } l = 0, 1, 2, \ldots \\ m \text{ for } L_z, &\quad \text{where } m = -l, -l+1, \ldots, l-1, \text{ or } l \end{aligned} \tag{5.9}$$

and then one of the two solutions of the differential equation (5.8) is free from infinities.

Strictly, a given eigenvalue $l(l+1)$ for $\boldsymbol{L}^2$ corresponds to the square of the angular momentum being equal to $\hbar^2 l(l+1)$. However, the usual terminology is to call the value assigned to l the *orbital angular momentum* of the state. The value of m is called the *magnetic quantum number*; the reason for this is that effects resulting from the quantisation of the z-component of the angular momentum are observed when atoms radiate photons in a magnetic field which points in the z-direction.

Notice that $[L_i, L_j] \neq 0$ for $i \neq j$. Hence two different components of $\boldsymbol{L}$ do not have simultaneous eigenfunctions, and we cannot simultaneously assign definite values to the corresponding observables. However, there is nothing special about the z-direction. Let L_α be the component of $\boldsymbol{L}$ in any other direction. Then it is easy to show that L_α commutes with $\boldsymbol{L}^2$. This can be done by direct calculation or by realising that *any* three orthogonal directions can be chosen for the coordinate axes, so that this new direction could equally well have been called the z-direction and the result follows because we already know that L_z commutes with $\boldsymbol{L}^2$. Instead of the simultaneous eigenfunctions Y_{lm} of L_z and $\boldsymbol{L}^2$, one can work with the analogous simultaneous eigenfunctions of L_α and $\boldsymbol{L}^2$, and the possible eigenvalues of L_α are m', where $m' = -l, -l+1, \ldots, l-1$ or l.

This means that it is possible to know simultaneously the exact values of both the orbital angular momentum l and of any one of its components. It is not possible however, to have certain knowledge simultaneously of more than one component of the angular momentum. (The only exception is when there is no orbital angular momentum, that is, $l = 0$. Then all components of the orbital angular momentum are zero.) Suppose that a measurement of the square of the orbital angular momentum gives the value $\hbar^2 l(l+1)$, and that a measurement of the z-component gives the value $m\hbar$, with m an integer between $-l$ and $+l$. If we now measure some other component, the value of l will remain unchanged but the system will have been disturbed in such a way that the information about L_z is lost. Instead, we will have determined a value m' for this new component, where m' also is some integer between $-l$ and $+l$. If we know the value of l and the original value of m, we can predict the probability of obtaining any of the possible values m', but we shall not describe this calculation here.

There is in quantum mechanics another type of angular momentum, apart from orbital angular momentum. This is an intrinsic angular momentum, not associated with the motion of a particle, and it is given the name *spin*. In quantum theory even a point particle may have spin. It turns out that in the case of spin angular momentum the restriction on the quantum number l is a little less severe: in addition to possibly being integral, it may also be half-integral. In particular, for the case of electrons and protons the spin angular momentum has $l = \frac{1}{2}$; we say that these particles have spin $\frac{1}{2}$, and this is always true for all electrons and all protons. The corresponding possible values for the magnetic quantum number associated with the spin are $m = \pm\frac{1}{2}$. (Photons, on the other hand, have spin 1.) In addition to their spin, the particles may carry orbital angular momentum, as we have already described. A full discussion of spin is beyond the scope of this book.

Spherically symmetric potentials

The properties of the orbital angular momentum are of particular importance in the analysis of the motion of a particle in any potential V that is spherically symmetric. If V is spherically symmetric about the origin, it depends only on the spherical polar coordinate r, and not on the angles θ and ϕ, so that $V = V(r)$. Then $[V, \boldsymbol{L}^2] = 0$, because $\boldsymbol{L}^2$ involves only the derivatives $\partial/\partial\theta$ and $\partial/\partial\phi$, but not $\partial/\partial r$ (see (5.3)). Similarly, $[V, L_z] = 0$. The operators $\boldsymbol{L}^2$ and L_z also commute with the kinetic-energy operator, because this operator is also spherically symmetric: $\hat{\boldsymbol{p}}^2/2M = (\hat{p}_x^2 + \hat{p}_y^2 + \hat{p}_z^2)/2M$ involves each of the three co-ordinate directions equally. (We have written the particle mass as M, to avoid confusion with the magnetic quantum number m.)

Hence for the case of a spherically symmetric potential

$$\begin{aligned}
[\boldsymbol{L}^2, L_z] &= 0 \\
[H, \boldsymbol{L}^2] &= 0 \qquad (5.10) \\
[H, L_z] &= 0
\end{aligned}$$

and we can find simultaneous eigenfunctions of H, $\boldsymbol{L}^2$ and L_z. That is, we can find stationary states where $\boldsymbol{L}^2$ and L_z take definite values, and we can use these values to label the stationary states. This result is not true in the case of a potential $V(\boldsymbol{r})$ that is not spherically symmetric, and that depends on the direction of $\boldsymbol{r}$ as well as on its magnitude.

Making use of (5.4), we have to solve the time-independent Schrödinger equation

$$-\frac{\hbar^2}{2M}\left(\frac{1}{r}\frac{\partial^2}{\partial r^2}r - \frac{1}{r^2}\boldsymbol{L}^2\right)\psi + V(r)\psi = E\psi. \qquad (5.11)$$

We look for solutions $\psi(\boldsymbol{r})$ that are simultaneous eigenfunctions of $\boldsymbol{L}^2$ and L_z, so that their angular dependence is

$$\psi(\boldsymbol{r}) = R(r)Y_{lm}(\theta, \phi), \qquad (5.12)$$

where Y_{lm} is defined by (5.6) and $R(r)$ is some function that remains to be found. Inserting this structure (5.12) into (5.11), and replacing $\boldsymbol{L}^2$

by its eigenvalue $l(l+1)$, we find that $R(r)$ satisfies the differential equation

$$\frac{-\hbar^2}{2Mr}\frac{d^2[rR(r)]}{dr^2}+\left(V(r)+\frac{\hbar^2 l(l+1)}{2Mr^2}\right)R(r)=ER(r). \qquad (5.13)$$

Because m does not appear in this equation, the eigenvalues E obtained from the equation will depend only on l and not on m. That is, for each value of l the states with the $(2l+1)$ possible different values of m given in (5.9) all have the same energy. We say that these states are $(2l+1)$-fold degenerate. The reason for this degeneracy is that, for a spherically symmetric potential, no direction in space is physically different from any other. For a rotationally symmetric Hamiltonian, the energy cannot depend on the magnitude of the angular-momentum component in any particular direction.

In the differential equation (5.13) for $R(r)$, the term involving $l(l+1)$ is simply added to $V(r)$. Thus it has the effect of an additional potential, which is repulsive. It is known as the *centrifugal potential*, by analogy with classical mechanics. Usually we expect to find that the ground state of a system (that is, the state of lowest energy, with the tightest binding) corresponds to $l=0$, so that the repulsive centrifugal potential is then absent. We shall see below in (5.17) that this is indeed the case for the hydrogen atom, though for some choices of the potential $V(r)$ the ground state can correspond to a larger value of l.

The hydrogen atom

The hydrogen atom consists of an electron of charge $-e$ bound to a proton of charge $+e$. Suppose that the centre of mass of the atom is at rest. Since the proton is nearly 2000 times heavier than the electron, the proton is then almost at rest and we may as a good approximation treat it as if it were. We explain how to relax this approximation later.

The approximation has reduced the problem to that of a single body, the electron, moving in the Coulomb field of the fixed proton. This field corresponds to the potential

$$V(r)=-e^2/4\pi\varepsilon_0 r, \qquad (5.14)$$

where we have taken the proton to be at rest at the origin. We insert this potential into (5.13) and look for a solution for the electron wave function that corresponds to a bound state. For a bound-state solution, we impose the condition that $R(r)\to 0$ as $r\to\infty$, so that the electron is localised somewhere near the proton. In a bound state, the electron does not have enough energy to escape to infinity. Because the potential (5.14) is zero at infinity, we therefore expect that the energy E will be negative, just as in the case of the one-dimensional square well (chapter 3). So we write

$$E = -\hbar^2\kappa^2/2M. \tag{5.15}$$

Then (5.13) becomes

$$\frac{-\mathrm{d}^2}{\mathrm{d}r^2}[rR(r)] + \left(\frac{l(l+1)}{r^2} - \frac{Me^2}{2\pi\varepsilon_0\hbar^2 r} + \kappa^2\right) rR(r) = 0. \tag{5.16}$$

We show in appendix A that this has solutions such that $R(r) \to 0$ at infinity and the behaviour at $r = 0$ is suitable if, and only if, $Me^2/4\pi\varepsilon_0\kappa\hbar^2$ is a positive integer n, and l is restricted to integers from 0 to $(n-1)$. That is,

$$E = \frac{-Me^4}{32\hbar^2\varepsilon_0^2\pi^2}\frac{1}{n^2} \qquad n = 1, 2, 3, \ldots$$
$$l = 0, 1, 2, \ldots, (n-1). \tag{5.17}$$

The integer n is known as the *principal quantum number.* The ground state corresponds to $n = 1$ and therefore $l = 0$.

It is interesting that the result (5.17) for E was first obtained by Bohr, using what is now known as 'old-fashioned' quantum theory. His work appeared some twelve years before the foundations of modern quantum theory were laid by Heisenberg, Schrödinger and Dirac. Bohr's theory turned out to be too simple and restrictive in its applications, but it does happen to give the correct result for the bound-state energies of the hydrogen atom. Bohr postulated that the electron moves round the proton in a classical circular orbit, and imposed a quantisation condition which requires that the orbital angular momentum be an integral multiple of $\hbar$.

The ground state of the atom corresponds to $n = 1$ and $l = 0$ in (5.17). When $l = 0$, we must have $m = 0$, and from (5.7) and (5.8) one can see that $Y_{00}(\theta, \phi)$ is a constant (remember that Y_{00} is the solution that is everywhere finite). We show in appendix A that the ground-state wave function is

$$(\pi a_0^3)^{-1/2}\,\mathrm{e}^{-r/a_0}$$

where

$$a_0 = 4\pi\hbar^2\varepsilon_0/Me^2. \tag{5.18}$$

It happens that a_0 is the radius of the ground-state orbit of the electron in Bohr's model; it is known as the *Bohr radius*. In the proper quantum theory the expectation value of the radius of the atom in its ground state is

$$\langle r \rangle = \int \mathrm{d}^3\boldsymbol{r}\, |\Psi|^2 r = 4\pi \int_0^\infty \mathrm{d}r\, r^3 \frac{\mathrm{e}^{-2r/a_0}}{\pi a_0^3} = \tfrac{3}{2} a_0.$$

Notice that in addition to the $(2l+1)$-fold degeneracy with respect to m, which is found for any spherically symmetric potential, we find in (5.17) an additional degeneracy. For each value of n, we have the same value of E for the n different values of l. This degeneracy is peculiar to the Coulomb potential, and so it is known as *accidental degeneracy*.

Many-electron atoms

Consider now an atom of atomic number $Z > 1$, so that the charge on the nucleus is Ze and there are Z electrons when the atom is electrically neutral. Each electron interacts with the nucleus through a potential $-Ze^2/4\pi\varepsilon_0 r$ so that if we neglect the interactions between the electrons the possible electron energy levels are obtained from (5.17) by making the replacement $e \to Z^{1/2}e$. The neglect of the electron–electron interaction turns out to be a fair approximation if Z is not too large.

As we have explained, the electron has spin $\frac{1}{2}$. A basic principle of quantum mechanics, the *Pauli exclusion principle*, states that no two identical particles of spin $\frac{1}{2}$ may occupy the same quantum-mechanical state. Corresponding to each wave function $R(r)Y_{lm}(\theta, \phi)$ there are two different electron states, because the z-component of the electron spin can be either $+\frac{1}{2}$ or $-\frac{1}{2}$. Hence in the ground state of the atom, two electrons occupy the lowest-energy level (5.17), that is, $n = 1$, $l = 0$. In the next level, $n = 2$, there can be two electrons with $l = 0$, and six with $l = 1$, because then m can take any of the three values $-1, 0, +1$. We fill in the levels in this way, starting from the lowest, until we have allocated all the electrons, and so obtain the configuration of the ground state of the atom, (For an atom with very many electrons, the discussion is more complicated, because the energy of the interactions between pairs of electrons cannot then be neglected.)

Two-body systems

For the hydrogen atom at rest, we assumed that it is a good approximation to regard the proton as being at rest. We then wrote the Schrödinger equation for the wave function $\psi(\boldsymbol{r})$ for the electron. If we do not make this approximation, which reduces the problem to the single-body case, we must introduce a wave function $\psi(\boldsymbol{r}_1, \boldsymbol{r}_2)$ that depends on the coordinate $\boldsymbol{r}_1$ of the electron and also on the coordinate $\boldsymbol{r}_2$ of the proton (see problem 2.5).

Generally, for a system of two interacting particles the potential will be a function of the separation between the particles. Hence the Hamiltonian operator takes the form

$$H = -(\hbar^2/2M_1)\nabla_1^2 - (\hbar^2/2M_2)\nabla_2^2 + V(\boldsymbol{r}_1 - \boldsymbol{r}_2), \tag{5.19}$$

where ∇_1^2 operates only on the coordinate $\boldsymbol{r}_1$ and ∇_2^2 operates only on the coordinate $\boldsymbol{r}_2$. In (5.19) the first two terms are the separate kinetic-energy operators for the two particles. It is convenient to change to centre-of-mass variables, as one would do in solving a two-body problem in classical mechanics:

$$\boldsymbol{r} = \boldsymbol{r}_1 - \boldsymbol{r}_2, \qquad \boldsymbol{R} = \frac{M_1\boldsymbol{r}_1 + M_2\boldsymbol{r}_2}{M_1 + M_2}, \tag{5.20}$$

so that $\boldsymbol{r}$ is the relative separation of the two particles and $\boldsymbol{R}$ is the position of their centre of mass. In terms of these variables, (5.19) may be reduced with a little algebra to the form

$$H = -(\hbar^2/2\mathcal{M})\nabla_{\boldsymbol{R}}^2 - (\hbar^2/2\mu)\nabla_{\boldsymbol{r}}^2 + V(\boldsymbol{r}), \tag{5.21}$$

where

$$\mathcal{M} = M_1 + M_2, \qquad \mu = M_1M_2/(M_1 + M_2). \tag{5.22}$$

The quantity μ is known as the *reduced mass* of the system.

Because V does not depend on $\boldsymbol{R}$, the time-independent Schrödinger equation

$$H\psi(\boldsymbol{r}_1, \boldsymbol{r}_2) = E_{\text{tot}}\psi(\boldsymbol{r}_1, \boldsymbol{r}_2)$$

has solutions of the form

$$\psi(\boldsymbol{r}_1, \boldsymbol{r}_2) = \psi_1(\boldsymbol{r})\psi_2(\boldsymbol{R}),$$

where

$$\left(\frac{-\hbar^2}{2\mu}\nabla_r^2 + V(\boldsymbol{r})\right)\psi_1(\boldsymbol{r}) = E\psi_1(\boldsymbol{r}),$$

$$\frac{-\hbar^2}{2\mathcal{M}}\nabla_{\boldsymbol{R}}^2\psi_2(\boldsymbol{R}) = E'\psi_2(\boldsymbol{R}), \tag{5.23}$$

$$E + E' = E_{\text{tot}}.$$

The second equation in (5.23) says that the centre of mass $\boldsymbol{R}$ of the system has a motion equivalent to that of a single particle of mass $\mathcal{M}$ moving freely, in zero potential. If the system is at rest, $E' = 0$. The first equation states that the relative motion of the two particles is the same as the motion of a single particle of mass μ moving in the potential V. This result is precisely that obtained in classical mechanics.

Hence in the expression (5.17) for the energy levels of the hydrogen atom, the mass M of the electron should be replaced by the reduced mass μ of the system. However, because the mass of the electron is much smaller (by a factor of about 2000) than that of the proton, $\mu \approx M$.

When the result (5.17), modified in this way, is compared with experiment, there is good agreement. As we explained in chapter 1, the comparison is made by examining the emission or absorption spectrum of the atom. The main discrepancy between theory and experiment may be accounted for by including in the Hamiltonian magnetic interaction terms in addition to the electrostatic Coulomb potential. The most important of these terms arises because the electron has both orbital and spin angular momentum. It can be shown from relativistic quantum mechanics that a charged particle of spin $\frac{1}{2}$, such as the electron, necessarily carries an intrinsic magnetic dipole moment, and the magnitude of this dipole moment can be calculated with high accuracy. The orbital motion of the electron generates a magnetic field; in rather crude terms this is because a charged particle moving in a closed orbit is equivalent to a current loop. The energy of interaction between this magnetic field and the electron's dipole moment must be included in the Hamiltonian, and it leads to a small correction to the energy levels (5.17). For a given value of n, the states with different values of l are then not exactly degenerate. There are other effects of a relativistic origin, and when all corrections have been made the calculations agree with experiment to better than one part in 10^6.

The deuteron

The nucleus of the isotope of hydrogen known as deuterium is called the deuteron. It is a bound state of a neutron and a proton. As their masses M are almost equal, the reduced mass of the system is $\mu \approx \frac{1}{2}M$.

According to (5.23), the relative motion of the proton and the neutron is described by the Schrödinger equation

$$[(-\hbar^2/2\mu)\nabla^2 + V]\psi(\boldsymbol{r}) = E\psi(\boldsymbol{r}). \tag{5.24}$$

Suppose that V depends only on $|\boldsymbol{r}| = r$, and treat this as in (5.11), where now $\boldsymbol{L}$ is associated with the *relative* orbital angular momentum of the two particles. Replace $\boldsymbol{L}^2$ by its eigenvalue, as in (5.13). We explained that we expect to find that the ground state of the system, with the tightest binding, corresponds to $l = 0$, so that the repulsive centrifugal potential is absent.

As a simple model for the potential, take the *spherical well*

$$\begin{aligned} V(r) &= -U \quad \text{for } r < a \\ &= 0 \quad\quad \text{for } r > a. \end{aligned} \tag{5.25}$$

Then for $l = 0$

$$\begin{aligned} (\mathrm{d}^2/\mathrm{d}r^2)rR(r) &= k^2 rR(r) \quad \text{for } r > a \\ (\mathrm{d}^2/\mathrm{d}r^2)rR(r) &= -K^2 rR(r) \quad \text{for } r < a, \end{aligned} \tag{5.26}$$

where $$E = -\hbar^2 k^2/2\mu, \qquad E + U = \hbar^2 K^2/2\mu. \tag{5.27}$$

Hence we have the solution

$$\begin{aligned} rR(r) &= A\,\mathrm{e}^{-kr} \quad \text{for } r > a \\ &= B \sin Kr \quad \text{for } r < a, \end{aligned} \tag{5.28}$$

where A and B are constants. Here we have imposed the constraint that $R(r) \to 0$ as $r \to \infty$, so that we indeed have a bound state. We have also required that $R(r)$ does not diverge at the origin: if it possessed a singularity of the form r^{-1}, it would not, in fact, satisfy the Schrödinger equation at the point $r = 0$. At $r = a$, we want both $R(r)$ and $R'(r)$ to be continuous. This gives the equation

$$k = -K \cot Ka \tag{5.29}$$

which, together with (5.27), corresponds to an equation for E with discrete solutions.

It turns out that when realistic values are taken for U and a, the deuteron is very weakly bound and the ground state is the only bound state. The wave function goes to zero very slowly as $r \to \infty$, and $\langle r \rangle \approx 2a$.

That is, most of the time the neutron and proton are separated by a distance greater than the range a of the potential. The possibility of this being so, and of the system nevertheless being a bound state, is a peculiarity of quantum mechanics.

Problems

5.1 How does the function $Y_{lm}(\theta, \phi)$ change if the coordinate axes are rotated through an angle α about the z-direction?

5.2 The quantity $e^2/4\pi\varepsilon_0\hbar c$ is called the *fine-structure constant.* Verify that it is dimensionless; it is approximately equal to $1/137$. Calculate also the Bohr radius, given in (5.18).

5.3 Calculate the energy levels of the hydrogen atom according to Bohr's theory as follows. Assume that the electron is in circular orbit and calculate its orbital angular momentum $\hbar L$ in classical mechanics. Then require L to be an integer. Verify that this leads to the same energy levels as the correct procedure based on Schrödinger's equation.

Calculate the velocity of the electron in the lowest-energy Bohr orbit and determine whether the use of non-relativistic kinematics is justified.

5.4 Derive an equation that determines the energies of the bound states of orbital angular momentum $l=0$ of a particle moving in the attractive potential

$$V(\mathbf{r})=(-\hbar^2U/2m)\delta(r-a),$$

where U and a are constant. [See the discussion of delta-function potentials at the end of chapter 3.]

5.5 The alkali atoms have an electronic structure that is approximately hydrogen-like, in that their chemical properties and spectral lines are determined essentially by a single electron. As a model for the potential in which this electron moves, take

$$V(r)=(-e^2/4\pi\varepsilon_0 r)(1+b/r).$$

Following the procedure described in appendix A, show that the energy levels are

$$E_{nl}=(-e^2/8\pi\varepsilon_0 a_0)[n-D(l)]^{-2},$$

where

$$D(l)=-[(l+\tfrac{1}{2})^2-2b/a_0]^{1/2}+l+\tfrac{1}{2}.$$

5.6 In a one-dimensional problem, two particles, each of mass m, interact through the potential $\frac{1}{2}m\omega^2(x_1-x_2)^2$, where x_1 and x_2 are their position coordinates. Find the energy levels of the system when its centre of mass is at rest.

Revision quiz

This quiz is designed to check that you have followed and understood the basic features of quantum theory. The answers to the questions may be found on the pages indicated.

1. What are the diameters of an atom and of its nucleus? (p. 1)
2. What do we learn from the photoelectric effect? (p. 3)
3. What is the wavelength associated with an electron, and how do we know this? (p. 6)
4. What are the operators corresponding to the momentum and energy of a particle? (pp. 9 and 10)
5. Write down the time-dependent Schrödinger equation. How and under what circumstances can one derive the time-independent Schrödinger equation from it? (p. 11)
6. What is the probability interpretation of the wave function? (p. 13)
7. What is the expression for the particle flux $\boldsymbol{j}$ in terms of the wave function? (p. 16)
8. What are the continuity conditions that are usually satisfied by the wave function, and how are they derived? (p. 16)
   Why are the conditions different for the problem of a particle in a box and for a particle moving in a delta-function potential? What are the conditions in these cases? (pp. 18 and 27)
9. What is the tunnel effect? (p. 24)
10. How does one use stationary-state solutions to express general solutions of the time-dependent Schrödinger equation? What mathematical assumption is implied? (p. 31)
11. What is the Heisenberg uncertainty principle? (p. 33)
    Under what conditions can one simultaneously know the values of two observables? (p. 38)
12. What is meant by a Hermitian operator? (p. 34)
    What properties do its eigenvalues and eigenfunctions have? (p. 35)
13. What are the basic assumptions about observables? (p. 37)
14. What is meant by the expectation value of an observable, and how is it calculated? (p. 38)
15. Why is the orbital-angular-momentum operator useful in problems where the potential is spherically symmetric? (p. 45)
16. Describe the energy levels of the hydrogen atom, and their degeneracies. (p. 47)
17. What is the Pauli exclusion principle? (p. 48)
18. How does one deal with a two-body system? (p. 49)

6

The hydrogen molecule

In chapter 5 we saw how in quantum mechanics electrons are bound to nuclei so as to form atoms. We now give a rather abbreviated account of how atoms bind together to form molecules. There is more than one type of molecular binding. We shall confine our discussion to the type known as *covalent binding*. The possibility of this type of binding relies on an effect that is peculiar to quantum mechanics, the tunnel effect, which we have already encountered in chapter 3.

The ionised hydrogen molecule

The simplest molecule is the ionised hydrogen molecule, which consists of two protons and one electron. The Coulomb force between the two protons tends to push them apart; we investigate how the presence of the electron overcomes this repulsion and holds the molecule together.

An exact calculation is difficult, but we can discuss the general features of the bonding by making suitable approximations. As the protons are much heavier than the electron, we may neglect their motion compared with that of the electron, and so regard them as fixed. We show that the expectation value of the energy, considered as a function of the proton separation R, has a minimum for a certain value of R, so that there is a stable equilibrium configuration.

Suppose first that R is so large that in the vicinity of each of the protons the Coulomb field of the other is completely negligible. Then there are two sets of stationary states: the electron is bound to one or other of the two protons with a wave function corresponding to the ordinary hydrogen atom. In particular, the lowest-energy level of the system is E_0, where E_0 is the energy of the ground state of the hydrogen atom. There are two different states corresponding to this energy. In one of them the electron is described by the wave function $\phi(\boldsymbol{r}-\boldsymbol{r}_1)$, where $\boldsymbol{r}_1$ is the coordinate vector of one of the protons and ϕ is the ground-state wave function (5.18) of the hydrogen atom:

$$\phi(\boldsymbol{r}-\boldsymbol{r}_1)=(\pi a_0^3)^{-1/2}\,\mathrm{e}^{-|\boldsymbol{r}-\boldsymbol{r}_1|/a_0}. \tag{6.1}$$

In the other state, the wave function is $\phi(\boldsymbol{r}-\boldsymbol{r}_2)$, where $\boldsymbol{r}_2$ is the coordinate vector of the other proton. Notice that these wave functions are real.

Now reduce R to a value such that the Coulomb field of one proton is beginning to be felt in the vicinity of the other. In classical physics the configuration of the system would be little altered, because it would require energy from outside to tear the electron from one proton and attach it to the other. But in quantum mechanics the tunnel effect does allow this transition to occur without any external energy being supplied. This means that $\phi(\boldsymbol{r}-\boldsymbol{r}_1)$ and $\phi(\boldsymbol{r}-\boldsymbol{r}_2)$ are no longer stationary-state wave functions. The system eventually settles down into a rather different stationary state, which we now discuss.

The stationary-state wave functions are the eigenfunctions of the Hamiltonian

$$H=-\frac{\hbar^2}{2m}\nabla_r^2-\frac{e^2}{4\pi\varepsilon_0|\boldsymbol{r}-\boldsymbol{r}_1|}-\frac{e^2}{4\pi\varepsilon_0|\boldsymbol{r}-\boldsymbol{r}_2|}+\frac{e^2}{4\pi\varepsilon_0 R}, \tag{6.2}$$

where m is the mass of the electron. Here, the first term is the kinetic-energy operator for the electron – remember that we are neglecting the kinetic energies of the protons. We write the stationary-state wave function for the ground state of the molecule as a superposition of $\phi(\boldsymbol{r}-\boldsymbol{r}_1)$ and $\phi(\boldsymbol{r}-\boldsymbol{r}_2)$:

$$\psi(\boldsymbol{r})=c_1\phi_1+c_2\phi_2$$

where

$$\phi_i=\phi(\boldsymbol{r}-\boldsymbol{r}_i)\qquad i=1,2. \tag{6.3}$$

Then $|c_1|^2$ is the probability of finding that the electron is in the configuration with wave function ϕ_1, that is, in orbit round the first proton. Similarly, $|c_2|^2$ is the probability of finding that it is in orbit round the second proton. The expansion (6.3) represents an approximation; it should contain also terms corresponding to configurations in which the electron is bound to one of the protons with a wave function associated with one of the excited states of the hydrogen atom. However, these terms are important only when R is taken to be so small that the extra Coulomb energy of repulsion between the protons is readily able to excite the electron into one of these states.

The wave function $\psi(\boldsymbol{r})$ has to satisfy the time-independent Schrödinger equation $H\psi = E\psi$. Insert in this equation the expansion (6.3) for ψ. Then pre-multiply the equation by ϕ_1 and integrate over all space, and repeat this with ϕ_2:

$$\begin{aligned} c_1(H_{11}-E) &= c_2(K_{12}E - H_{12}) \\ c_1(K_{12}E - H_{21}) &= c_2(H_{22}-E), \end{aligned} \tag{6.4}$$

where

$$\begin{aligned} H_{ij} &= (\phi_i, H\phi_j) \qquad i, j = 1 \text{ or } 2 \\ K_{12} &= (\phi_1, \phi_2) = (\phi_2, \phi_1) \end{aligned} \tag{6.5}$$

and we have used the normalisation $(\phi_1, \phi_1) = (\phi_2, \phi_2) = 1$ implied by (6.1). Because of the symmetry of the problem, $H_{11} = H_{22}$ and $H_{12} = H_{21}$. This means that eliminating c_1 and c_2 from the equations (6.4) gives

$$(H_{11}-E)^2 = (K_{12}E - H_{12})^2$$

which leads to the two solutions

$$\begin{aligned} &\text{(i)} \qquad E = E_+ = \frac{H_{11}+H_{12}}{1+K_{12}}, \qquad \text{with } c_1 = c_2 \\ &\text{(ii)} \qquad E = E_- = \frac{H_{11}-H_{12}}{1-K_{12}}, \qquad \text{with } c_1 = -c_2. \end{aligned} \tag{6.6}$$

In both these solutions, $|c_1|^2 = |c_2|^2$, that is, there is *equal* probability of finding the electron in orbit round either proton. This result shows how dramatic are the consequences of the tunnel effect; the situation is very different from that in classical mechanics. Crudely speaking, the electron continually tunnels back and forth from the potential well surrounding one proton to that surrounding the other.

We must now attempt a rough calculation of H_{11} and H_{12}. The first two terms of the Hamiltonian H in (6.2) are nothing but the hydrogen-atom Hamiltonian of which ϕ_1 is the ground-state eigenfunction, so

$$H_{11} = E_0 - \frac{e^2}{4\pi\varepsilon_0}\int d^3\boldsymbol{r}\frac{\phi_1^2}{|\boldsymbol{r}-\boldsymbol{r}_2|} + \frac{e^2}{4\pi\varepsilon_0 R}. \tag{6.7}$$

Because of the definition in (6.3) of ϕ_1, we may write the integral in (6.7) as

$$-\frac{e^2}{4\pi\varepsilon_0}\int d^3\boldsymbol{r}'\frac{[\phi(\boldsymbol{r}')]^2}{[(\boldsymbol{r}'-\boldsymbol{R})^2]^{1/2}}, \tag{6.8}$$

where we have written $\boldsymbol{r}'=\boldsymbol{r}-\boldsymbol{r}_1$ and $\boldsymbol{R}=\boldsymbol{r}_2-\boldsymbol{r}_1$. We see from (6.1) that $\phi(\boldsymbol{r}')$ is very small when $|\boldsymbol{r}'|\gg a_0$. So if $R\gg a_0$, then $|\boldsymbol{r}'|\ll R$ throughout that part of the integration from which the integral receives most of its contribution. Hence we may use the expansion

$$\begin{aligned}1/[(\boldsymbol{r}'-\boldsymbol{R})^2]^{1/2} &= [r'^2-2\boldsymbol{r}'\cdot\boldsymbol{R}+\boldsymbol{R}^2]^{-1/2}\\ &= (1/R)[1+(\boldsymbol{r}'\cdot\boldsymbol{R}/R^2)+\cdots].\end{aligned} \tag{6.9}$$

When we insert this expansion into the integral, the first term integrates to give $-e^2/4\pi\varepsilon_0 R$. The second term, being odd in each of the three coordinates of $\boldsymbol{r}'$, integrates to give zero. Hence for large R, (6.7) reduces to

$$H_{11}\approx E_0. \tag{6.10a}$$

This approximation is correct up to and including terms of order a_0/R^2, where the factor a_0 appears here because ϕ_1 is appreciable only in the region where $r'\ll a_0$. That is, the correction is of order a_0^2/R^3.

On the other hand, when $R\to 0$ the integral (6.8) evidently remains finite. The last term in the Hamiltonian (6.7) diverges as $R\to 0$, and so dominates for small R. Hence for small R we have

$$H_{11}\approx e^2/4\pi\varepsilon_0 R. \tag{6.10b}$$

We consider H_{12} similarly:

$$H_{12}=K_{12}E_0-\frac{e^2}{4\pi\varepsilon_0}\int d^3\boldsymbol{r}\frac{\phi_1\phi_2}{|\boldsymbol{r}-\boldsymbol{r}_1|}+\frac{K_{12}e^2}{4\pi\varepsilon_0 R}. \tag{6.11}$$

The integral here is $-V(R)$, where

$$V(R)=\frac{e^2}{4\pi\varepsilon_0}\int d^3\boldsymbol{r}'\frac{\phi(\boldsymbol{r}')\phi(\boldsymbol{r}'-\boldsymbol{R})}{r'}. \tag{6.12}$$

This integral does not diverge at $r'=0$, because the volume element $d^3\boldsymbol{r}'$ contains a factor r'^2 when it is written in spherical polar coordinates. Nevertheless, relatively small values of r' are most important in the integral, and so when R is large

$$V(R)\gg\frac{e^2}{4\pi\varepsilon_0 R}\int d^3\boldsymbol{r}'\phi(\boldsymbol{r}')\phi(\boldsymbol{r}'-\boldsymbol{R})=\frac{e^2}{4\pi\varepsilon_0 R}K_{12}.$$

Hence for large R we may neglect the last term in (6.11) compared with the others, and

$$H_{12} \approx K_{12}E_0 - V(R). \tag{6.13a}$$

The integral (6.12) is evidently finite at $R = 0$, so that for small R the last term in (6.11) is dominant and

$$H_{12} \approx K_{12}\, e^2/4\pi\varepsilon_0 R. \tag{6.13b}$$

Finally, from its definition in (6.5), K_{12} is small for large R, because when ϕ_1 is large ϕ_2 is nearly zero, and vice versa. If we use this, and put (6.10) and (6.13) together, the energies $E_\pm$ in (6.6) become

$$E_\pm \approx \begin{cases} E_0 \mp V(R) & R \text{ large} \\ e^2/4\pi\varepsilon_0 R & R \text{ small.} \end{cases} \tag{6.14}$$

By inspection of (6.12) we see that $V(R)$ is positive, and it decreases in magnitude as R is increased, becoming small for large R. Hence the plots of $E_\pm$ against R must have the shapes drawn in figure 6.1. Because of (6.14), E_+ and E_- are both very large and positive for small R. At large R, E_+ approaches its asymptotic value E_0 from below. Therefore E_+ must have a minimum at some value of R; that is, there is a configuration of stable equilibrium. The corresponding value of the energy E_+ is less than E_0, which is the minimum energy of a separate proton and hydrogen atom, so that the configuration is a bound state.

Other molecules

A similar type of covalent binding provides most of the binding in the neutral hydrogen molecule, in most organic molecules, and in some inorganic molecules. (However, most inorganic molecules are bound

Figure 6.1. $E_\pm$ plotted against R.

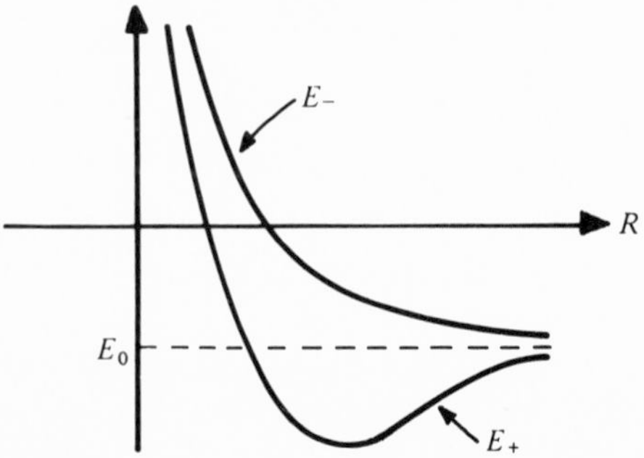

mainly by a different mechanism, known as ionic binding.) When only one electron is involved, the covalent bond can be strong only between similar atoms: the symmetry properties $H_{11}=H_{22}$ and $H_{12}=H_{21}$ were of crucial importance in our analysis. (See problem 6.3.) But two electrons can form a strong covalent bond between two atoms A and B even when these are different. The theory is similar, but with ϕ_1 now the wave function for the state where the first electron is attached to atom A and the second to atom B, and ϕ_2 the wave function for the corresponding state where the electrons are interchanged. It is evident that in this case the equalities $H_{11}=H_{22}$ and $H_{12}=H_{21}$ do hold.

Problems

6.1 As a one-dimensional model for the ionised hydrogen molecule, suppose that the potential near an atom is a delta function, so that for the molecule

$$V(x)=(-\hbar^2U/2m)[\delta(x-R)+\delta(x+R)].$$

Show that the wave function of the electron has the form

$$\psi(x)=\begin{cases}A(e^{\kappa x}\pm e^{-\kappa x}) & 0\leqslant x<R\\ B\,e^{-\kappa x} & x>R,\end{cases}$$

and find equations that give the bound-state energy as a function of R. [You may find it helpful to refer to the discussion of delta-function potentials at the end of chapter 3.]

6.2 The muon is a particle whose properties are closely similar to those of the electron, except that it is some 200 times heavier (and it is unstable, with a lifetime of about 10^{-6} s). If it is used to bind together two nuclei, how does the size of the resulting molecule compare with that of the electron-bound molecule? [We explained at the end of chapter 5 that the radius of the deuteron, the nucleus of deuterium, is quite large. When a muon binds two deuterons together the distance between the deuterons is sufficiently small for there to be a substantial overlap between their wave functions, and so there is a good chance that the two nuclei will fuse and form a helium nucleus before the muon decays.]

6.3 Suppose that, in (6.4), $H_{11}\neq H_{22}$ and $H_{12}\neq H_{21}$. Let the energies of the ground states of the two separate atoms be respectively E_{01} and E_{02}, with $E_{01}<E_{02}$. Show that for large R the lowest-energy eigenvalue E differs from E_{01} by an amount that is not linear in small quantities, as in (6.14), but quadratic. [The consequence is that a single electron does not give strong covalent binding between unlike atoms.]

6.4 Three atomic nuclei are fixed at the corners of an equilateral triangle and an electron is introduced into the system. If this electron were bound to the atom i in isolation, it would have wave function ϕ_i. Obtain expressions analogous to (6.6) for the energy levels.

7

Introduction to perturbation theory

In this chapter we describe the approximation method for problem solving known as perturbation theory. There are two types of perturbation theory, time-independent perturbation theory, used to find the stationary states of a system, and time-dependent perturbation theory, used to calculate certain quantities called transition probabilities.

Time-independent perturbation theory

In chapter 5 we described a model of the deuteron in which the proton and neutron are bound together by the simple spherical-well potential given in (5.25). This is certainly not the true potential, but it is the model that most easily leads to a solution of the time-independent Schrödinger equation. The true potential will also be an attractive potential, and so the simple form (5.25) is an approximation to it. Thus the question arises: suppose that we are able to find the eigenvalues of a Hamiltonian H_0, can we now calculate, at least approximately, the eigenvalues of another Hamiltonian

$$H = H_0 + H_1 \tag{7.1}$$

that is not too different from H_0?

Let $\phi_j(\boldsymbol{r})$, $j = 1, 2, 3, \ldots$ be the various eigenfunctions of H_0, corresponding to eigenvalues E_j. Assume that there is no degeneracy, that is, that the E_j are all distinct. We shall also suppose that the ϕ_j are orthonormal, meaning that they meet the condition

$$(\phi_j, \phi_k) = \delta_{jk}, \tag{7.2}$$

and that they form a complete set, as discussed in chapter 4. This allows the eigenfunctions ψ_i of H to be expanded as linear combinations of the ϕ_j:

$$\psi_i = \sum_j c_{ij}\phi_j, \tag{7.3}$$

with the c_{ij} constant.

We consider the case where H_1 is small. By this we mean that corresponding to a given eigenfunction ϕ_i of H_0 with eigenvalue E_i, there is an eigenfunction ψ_i of H that is not too different from ϕ_i, and that corresponds to an eigenvalue $E_i + \Delta E_i$ that is not too different from E_i. More precisely, we expect that our approximations will be good when ΔE_i is small compared with the difference between E_i and the next eigenvalue of H_0, and that all the coefficients c_{ij}, except for c_{ii}, are such that $|c_{ij}| \ll 1$. Then if we normalise ψ_i so that $(\psi_i, \psi_i) = 1$, which implies that

$$\sum_j |c_{ij}|^2 = 1, \tag{7.4}$$

c_{ii} will differ from 1 by a quadratically small quantity.

On inserting the expansion (7.3) into the time-independent Schrödinger equation $H\psi_i = (E_i + \Delta E_i)\psi_i$, we obtain

$$(H_0 + H_1) \sum_j c_{ij}\phi_j(\boldsymbol{r}) = (E_i + \Delta E_i) \sum_j c_{ij}\phi_j(\boldsymbol{r}). \tag{7.5}$$

To simplify this equation, use the definition $H_0\phi_j = E_j\phi_j$ of the eigenfunctions ϕ_j. Also, neglect terms that are quadratic in small quantities; that is, omit those terms where either H_1 or ΔE_i is multiplied by c_{ij} with $i \neq j$, and replace c_{ii} by 1. Then

$$H_1\phi_i(\boldsymbol{r}) + \sum_{j \neq i} (E_j - E_i)c_{ij}\phi_j(\boldsymbol{r}) = \Delta E_i\phi_i(\boldsymbol{r}). \tag{7.6}$$

Now pre-multiply by $\phi_i^*(\boldsymbol{r})$ (remember that H_1 may contain differential operators), integrate over all space and use the orthonormality condition (7.2). This gives

$$\Delta E_i = (\phi_i, H_1\phi_i). \tag{7.7}$$

That is, the shift ΔE_i in the level E_i resulting from the addition of the perturbation H_1 to the original Hamiltonian H_0, is just the expectation value of H_1 calculated from the original wave function ϕ_i.

Usually this is as far as we need go in the calculation: the level shift ΔE_i is the quantity of most interest. But if we wish also to calculate the perturbed wave function ψ_i, we instead pre-multiply (7.6) by $\phi_k^*(\boldsymbol{r})$, with $k \neq i$, and again integrate over all space. We then find that

$$c_{ik} = (\phi_k, H_1\phi_i)/(E_i - E_k) \qquad k \neq i. \tag{7.8}$$

By inserting this result into (7.3), together with $c_{ii} = 1$, we have an expression for $\psi_i(\boldsymbol{r})$.

Our calculations are correct to first order in small quantities. It is possible to extend the calculation to give expressions for the higher-order terms, but we shall not do this here.

Notice the importance of our original assumption that the level E_i is non-degenerate; otherwise the expression (7.8) that we have obtained for c_{ik} would be singular for some k, and our assumption that all the c_{ik} with $k \neq i$ are small would be inconsistent. If we are interested in the perturbation ΔE_i to a particular level E_i, our calculation is valid so long as that level is non-degenerate; it does not matter if some of the other levels happen to be degenerate. It is possible also to deal with the case where the level of interest is degenerate. For two degenerate states i and k such that the numerator in (7.8) vanishes, there is no problem, and the previous analysis applies without modification. We shall not pursue the case where the numerator does not vanish.

Example

If we want to apply perturbation theory to calculate the eigenvalues of a Hamiltonian H, the first step is to write it in the form (7.1). This division of H into two parts H_0 and H_1 may be a purely mathematical procedure; the only requirements are that we can find the eigenstates of H_0, and that H_1 is small in the sense that we have described. Usually, however, the way in which we choose to divide H will have an obvious physical significance.

As a simple example, consider again the problem of a particle in a rectangular box, as described at the beginning of chapter 3. Suppose that the particle does not now move freely within the box, but rather that it carries a charge e and is acted on by a uniform electric field $\mathscr{E}$. Take the field to be parallel to the x-axis, so that the potential takes the form

$$V = -e\mathscr{E}x. \tag{7.9}$$

If either e or $\mathscr{E}$ is small, so that this potential is weak, an obvious choice is to identify H_1 with V, so that H_0 is the kinetic-energy operator whose eigenfunctions are given in (3.4), namely

$$\phi_{qrs}(\boldsymbol{r}) = \left(\frac{8}{abc}\right)^{1/2} \sin\frac{q\pi x}{a} \sin\frac{r\pi y}{b} \sin\frac{s\pi z}{c} \tag{7.10}$$

with q, r, s integers. The corresponding unperturbed energy levels are, from (3.5),

$$E_{qrs} = \frac{\hbar^2\pi^2}{2m}\left(\frac{q^2}{a^2}+\frac{r^2}{b^2}+\frac{s^2}{c^2}\right). \tag{7.11}$$

In order to be able to apply the perturbation theory developed above, we must evidently confine ourselves to the case where the lengths a, b, c of the sides of the box are unequal, so as to ensure that the level E_{qrs}, for which the shift ΔE_{qrs} is being calculated, is non-degenerate.

From (7.7), the energy levels in the presence of the electric field are $E_{qrs} + \Delta E_{qrs}$, where

$$\Delta E_{qrs} = (\phi_{qrs}, -e\mathscr{E}x\phi_{qrs}) = -e\mathscr{E}\int_{\text{box}} \mathrm{d}x\, \mathrm{d}y\, \mathrm{d}z\, x\phi_{qrs}^2 = -\tfrac{1}{2}e\mathscr{E}a. \tag{7.12}$$

In this simple problem, the perturbation has turned out to be the same for every level.

Time-dependent perturbation theory

We now investigate the behaviour of a system under the influence of a perturbation that varies with time:

$$H = H_0 + H_1(t). \tag{7.13}$$

We suppose that, as before, H_0 does not vary with t. We again use the stationary-state wave functions of H_0 to make an expansion of the wave function $\Psi(\boldsymbol{r}, t)$ of the system. Because we are interested in the time dependence of Ψ, we must now include in the stationary-state wave functions of H_0 their time variation:

$$\Phi_j(\boldsymbol{r}, t) = \phi_j(\boldsymbol{r})\, \mathrm{e}^{-\mathrm{i}E_jt/\hbar}, \tag{7.14}$$

where the $\phi_j(\boldsymbol{r})$ are the same eigenfunctions as in (7.3). Because H varies with t, it does not have stationary-state wave functions; we must solve the time-dependent Schrödinger equation

$$H\Psi(\boldsymbol{r}, t) = \mathrm{i}\hbar\dot{\Psi}(\boldsymbol{r}, t). \tag{7.15}$$

Suppose that as $t \to -\infty$, $H_1(t) \to 0$, so that initially H coincides with

H_0. Thus initially the system may be in a stationary state of H_0. We shall suppose that it is initially in the stationary state i and we shall denote by Ψ_i the wave function that coincides with Φ_i at $t=-\infty$.

When the perturbation $H_1(t)$ begins to switch on, Ψ_i will start to become different from Φ_i. We use the expansion

$$\Psi_i(\boldsymbol{r}, t)=\sum_j a_{ij}(t)\Phi_j(\boldsymbol{r}, t). \tag{7.16}$$

Notice that the coefficients a_{ij} are functions of t, because the Φ_j are not stationary-state wave functions for the perturbed system. At $t=-\infty$, $a_{ii}=1$, and all the other a_{ij} are zero. Inserting this expansion into the Schrödinger equation (7.15), and using the equation $H_0\Phi_j(\boldsymbol{r}, t)=\mathrm{i}\hbar\dot{\Phi}_j$ which defines Φ_j, we find that

$$\mathrm{i}\hbar\sum_j \dot{a}_{ij}\Phi_j=\sum_j a_{ij}H_1\Phi_j. \tag{7.17}$$

On pre-multiplying this equation by Φ_k^*, and integrating over all space, we obtain, with the help of the orthonormality conditions $(\Phi_k, \Phi_j)=\delta_{kj}$,

$$\mathrm{i}\hbar\dot{a}_{ik}=\sum_j a_{ij}(\Phi_k, H_1\Phi_j). \tag{7.18}$$

Notice that in (7.17) and (7.18) we have had to be careful about the order in which we have written the wave functions Φ relative to H_1, because H_1 may contain the derivative operator $\partial/\partial\boldsymbol{r}$. However, we do not allow H_1 to contain the operator $\partial/\partial t$, so that H does not operate on the coefficient functions $a_{ij}(t)$; these simply multiply H.

So far, we have made no approximations: (7.18) is exactly equivalent to the time-dependent Schrödinger equation. We now suppose that H_1 is small. This means that we expect the coefficient functions $a_{ij}(t)$ to depart rather little from their original values, at least if the time t is not too large. We shall suppose that the value of t is such that $a_{ii}(t)$ is still close to 1, and that $a_{ij}(t)$ for $i\neq j$ is still small. We then retain in (7.18) only those terms that are first order in small quantities, which means that on the right-hand side we keep only the term in the sum for which $j=i$ and in it we replace a_{ii} by 1:

$$\mathrm{i}\hbar\dot{a}_{ik}(t)=(\Phi_k, H_1\Phi_i)=(\phi_k, H_1\phi_i)\,\mathrm{e}^{\mathrm{i}(E_k-E_i)t/\hbar}. \tag{7.19}$$

Integrating with respect to t, we have

$$a_{ik}(t)=\delta_{ik}-\frac{\mathrm{i}}{\hbar}\int_{-\infty}^{t}\mathrm{d}t'(\phi_k, H_1(t')\phi_i)\,\mathrm{e}^{\mathrm{i}(E_k-E_i)t'/\hbar}, \tag{7.20}$$

where the δ_{ik} is the integration constant that takes account of the initial conditions at $t=-\infty$.

Transition probability

We have supposed that initially, at $t=-\infty$, the Hamiltonian is just H_0, and that the system is then in the stationary state Φ_i of H_0. Suppose that after some time T the perturbation $H_1(t)$ vanishes, so that the Hamiltonian is again H_0. If we now measure the energy of the system, the system will be thrown into a stationary state of H_0 and will remain in this state. The probability that the measurement results in the value E_k, and so throws the system into the stationary state Φ_k, is, according to the discussion given in chapter 4 and to (7.16), just $|a_{ik}|^2$. Provided that this probability is small, so that our approximations are valid, we may calculate it from (7.20):

$$|a_{ik}(t)|^2=\frac{1}{\hbar^2}\left|\int_{-\infty}^{T} \mathrm{d}t'(\phi_k, H_1(t')\phi_i)\,\mathrm{e}^{\mathrm{i}(E_k-E_i)t'/\hbar}\right|^2 \qquad (k\neq i,\, t>T). \tag{7.21}$$

This, then, is the probability that our measurement finds that the perturbation H_1 has induced a transition from the state Φ_i to the state Φ_k. Notice that a_{ik} involves the term $(\phi_k, H_1\phi_i)$. Frequently one can show without detailed calculation that this term vanishes for all but a few final states ϕ_k, regardless of the value of t'. That is, without detailed calculation one can find those final states ϕ_k that are readily accessible from the initial state ϕ_i. Those final states ϕ_k for which (7.21) vanishes may still be possible final states, but with a transition probability that is non-zero only at second or higher order in the perturbation H_1. Such transitions are known as *forbidden transitions*; usually (though not always) their corresponding transition probabilities are very small.

Example

As an example, consider again the problem of a charged particle in a box, and let the perturbation again be caused by a uniform electric field in the x-direction, as given in (7.9). Suppose that this perturbation is switched on at time $t=0$, and then switched off again at time T. We assume that for $t<0$ the system is in the stationary state ϕ_{qrs}. The probability that a measurement at some time $t>T$ will find that there has been a transition to the stationary state ϕ_{QRS} is

$$(1/\hbar^2)\left|\int_0^T dt'(\phi_{QRS}, -e\mathscr{E}x\phi_{qrs})\, e^{i(E_{QRS}-E_{qrs})t'/\hbar}\right|^2$$
$$= (e^2\mathscr{E}^2/\hbar^2)|I(\omega)|^2(\phi_{QRS}, x\phi_{qrs}), \tag{7.22}$$

where

$$I(\omega) = \int_0^T dt'\, e^{i\omega t'}$$
$$\omega = (E_{QRS} - E_{qrs})/\hbar. \tag{7.23}$$

The term $(\phi_{QRS}, x\phi_{qrs})$ is readily evaluated, using the explicit form of the wave functions given in (7.10):

$$(\phi_{QRS}, x\phi_{qrs}) = \int_{\text{box}} dx\, dy\, dz \phi_{QRS}\, x\, \phi_{qrs}$$
$$= \delta_{Rr}\delta_{Ss} \times \begin{cases} 0 & Q-q \text{ even } (Q \neq q) \\ -8aQq/[\pi^2(Q^2-q^2)^2] & Q-q \text{ odd.} \end{cases} \tag{7.24}$$

So first-order perturbation theory allows only those transitions for which $R=r$, $S=s$ and Q differs from q by an odd integer. Alternatively, the system may remain in its initial state.

From (7.23) we find that

$$|I(\omega)|^2 = (4/\omega^2)\sin^2 \tfrac{1}{2}\omega T. \tag{7.25}$$

A plot of this function against ω is shown in figure 7.1. The height of the central peak is T^2, while that of the peaks next to it on either side is

Figure 7.1. Plot of $|I(\omega)|^2$ against ω (see (7.25)).

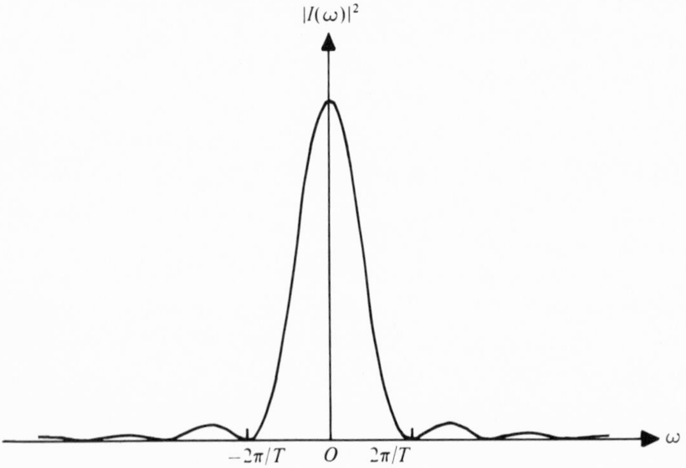

some twenty times less. Hence the transitions that occur most readily are those to levels for which ω lies within the central peak. This peak is centred on $\omega = 0$ and it has half-width $2\pi/T$. So the levels to which transitions are most likely are those whose energy E_{QRS} differs from the original energy E_{qrs} by an amount $\hbar\omega$ of order $\hbar/T$.

The shorter the duration T of the perturbation, the greater will be the spread in energy of this band of preferred final levels. This somewhat surprising result is peculiar to quantum mechanics, and is a manifestation of the *energy uncertainty principle.* We explained in chapter 4 that making a measurement on a system is liable to induce a perturbation, and, in fact, this principle also governs the accuracy with which it is possible to measure the energy of a system: it can be shown that if a limited time T is available for making the measurement, then the energy cannot be determined better than to within an amount of order $\hbar/T$. In particular, if a system is unstable and so has a finite lifetime τ, *no* measurement can determine its energy to within an accuracy better than about $\hbar/\tau$. The energy uncertainty principle is important also for understanding how the tunnel effect is possible: it allows a particle momentarily to acquire enough energy to surmount a potential barrier.

Sudden change in the Hamiltonian

Up to now, we have considered perturbations $H_1(t)$ that vanish for very large t, so that the Hamiltonian ultimately returns to its original form H_0. Consider now the case where the Hamiltonian is initially equal to the time-independent operator H_0, as before, but then at time $t = t_0$ the Hamiltonian instantaneously changes, so that for all $t > t_0$ it is equal to the time-independent operator H'_0. We suppose that we are able to solve the time-independent Schrödinger equation for both Hamiltonians, and that we know that the eigenvalues of H_0 are E_j, $j = 1, 2, \ldots$, and those of H'_0 are E'_j, $j = 1, 2, \ldots$ Let the system initially be in a stationary state of H_0, with energy E_i. We calculate the probability that a measurement of the energy at some time t later than t_0 will result in the value E'_j, and so throw the system into the corresponding stationary state of H'_0. In the previous work we calculated the first term in a power-series expansion for the probability; here we are able to give an exact expression.

Let the stationary states of H_0 have the orthonormal wave functions $\Phi_j(\boldsymbol{r}, t)$ as in (7.14). Let those of H_0' have the orthonormal wave functions

$$\Phi_j'(\boldsymbol{r}, t) = \phi_j'(\boldsymbol{r})\,\mathrm{e}^{-\mathrm{i}E_j't/\hbar} \tag{7.26}$$

where

$$(\phi_j', \phi_k') = \delta_{jk}.$$

Note that here the prime does not denote differentiation, and the functions Φ_j' may be completely different from the Φ_j. *We do not need to assume that the difference between H_0 and H_0' is small.* For $t < t_0$, the wave function $\Psi(\boldsymbol{r}, t)$ of the system is, by assumption, equal to Φ_i. For $t > t_0$ we expand it as a linear combination of the Φ_j':

$$\Psi(\boldsymbol{r}, t) = \begin{cases} \Phi_i(\boldsymbol{r}, t) & t < t_0 \\ \sum_j b_j \Phi_j'(\boldsymbol{r}, t) & t > t_0. \end{cases} \tag{7.27}$$

For $t > t_0$ the Φ_j' are stationary-state wave functions of the Hamiltonian, and therefore the coefficients b_j are constant. The wave function (7.27) will automatically satisfy the Schrödinger equation

$$H\Psi = \mathrm{i}\hbar\, \partial\Psi/\partial t \tag{7.28}$$

for both $t < t_0$ and $t > t_0$. At $t = t_0$, Ψ must be continuous. For if it were not, $\partial\Psi/\partial t$ would have a singularity proportional to $\delta(t - t_0)$ at $t = t_0$ (see appendix B), and the Schrödinger equation (7.28) would not then be satisfied at $t = t_0$ since H merely has a finite discontinuity there. The continuity condition at $t = t_0$ gives

$$\sum_j b_j \Phi_j'(\boldsymbol{r}, t_0) = \Phi_i(\boldsymbol{r}, t_0). \tag{7.29}$$

Multiplying this equation by $\Phi_k'^*(\boldsymbol{r}, t_0)$, integrating over all space and using the orthonormality conditions in (7.26), we obtain

$$b_k = (\phi_k', \phi_i)\,\mathrm{e}^{\mathrm{i}(E_k' - E_i)t/\hbar}. \tag{7.30}$$

Hence the probability that a measurement of the energy at some time $t > t_0$ yields the value E_k' is

$$|b_k|^2 = |(\phi_k', \phi_i)|^2. \tag{7.31}$$

Example: decay of tritium

Tritium is the isotope of hydrogen whose nucleus consists of one proton and two neutrons. When a neutron is not bound into a nucleus it is unstable: it undergoes the decay

$$\text{neutron} \rightarrow \text{proton} + \text{electron} + \text{neutrino}. \tag{7.32}$$

(The neutrino is a particle that has zero charge and also zero mass, so that like the photon it moves with velocity c.) Consider this decay in the frame of reference in which the neutron is at rest. The reason that the decay is possible is that the rest mass m_n of the neutron is greater than the sum of the rest masses of the proton and the electron, so that the initial-state energy $m_n c^2$ is sufficient both to create the final-state particles and to give them some kinetic energy. However, when the neutron is bound in a nucleus, the decay may not be possible. Consider, for example, the deuteron. If its neutron were to decay, the nucleus would have to break up, because there is no bound state of two protons. But this break-up is not allowed by energy conservation: the binding of the deuteron is strong enough to ensure that the rest mass of the deuteron is less than the sum of the rest masses of two protons and an electron.

In the case of tritium, energy conservation does allow one of its neutrons to decay as in (7.32). This is because there is a bound state of two protons and a neutron, namely the isotope ^{3}He of helium. The decay is

$$\text{tritium} \rightarrow {}^3\text{He} + \text{electron} + \text{neutrino}. \tag{7.33}$$

Further, the nucleus ^{3}He is bound rather more tightly than the original tritium nucleus, so that quite a lot of kinetic energy is available in the final state. The electron is so energetic that it rapidly escapes, rather than being trapped in an orbit corresponding to one of the atomic states of the helium atom.

Suppose now that, before its decay, the tritium nucleus was accompanied by an orbiting electron in the atomic ground state, so that the tritium atom was neutral. When the decay (7.33) occurs, the charge on the nucleus suddenly changes from e to $2e$, so that the stationary atomic states correspond to new electron wave functions. Because the ^{3}He has only one orbiting electron, it is not neutral but positively ionised. If we measure the energy of the orbiting electron after the decay of the nucleus, we do not necessarily find that the orbiting electron is in the new ground state. The probability that we do find this to be so is calculated from (7.31) by using the original ground-state wave function for ϕ_i and the new ground-state wave function for ϕ'_k.

According to (5.18) and the subsequent work in chapter 5,

$$\begin{aligned}\phi_i(\boldsymbol{r}) &= \pi^{-1/2}a_0^{-3/2}\,\mathrm{e}^{-r/a_0}\\ \phi'_k(\boldsymbol{r}) &= \pi^{-1/2}(\tfrac{1}{2}a_0)^{-3/2}\,\mathrm{e}^{-2r/a_0},\end{aligned} \tag{7.34}$$

where the Bohr radius a_0 is calculated in terms of the appropriate reduced mass and is approximately the same in both wave functions. Thus,

$$\begin{aligned}|(\phi'_k, \phi_i)|^2 &= \left|\int \mathrm{d}^3\boldsymbol{r}\,\phi'_k(\boldsymbol{r})\phi_i(\boldsymbol{r})\right|^2\\ &= (8/\pi^2 a_0^6)\left|\int 4\pi r^2\,\mathrm{d}r\,\mathrm{e}^{-3r/a_0}\right|^2\\ &= 512/699.\end{aligned} \tag{7.35}$$

So the probability that the helium ion is found to be in its ground state is close to 3/4.

If the helium atom is not in its ground state, and if the electron has not escaped from the atom, it will be in an excited state. In this case it will subsequently decay back into the ground state, emitting one or more photons in the process (see the next chapter). Hence tritium gas provides a useful source of illumination that needs no external supply of power.

Problems

7.1 A hydrogen atom is placed in a weak electric field $\mathscr{E}$. Show that to first order in $\mathscr{E}$ the ground-state energy is unchanged.

7.2 Calculate the first-order perturbation to the ground-state energy of a linear harmonic oscillator due to an additional potential εx^2. Verify that your result agrees with the exact answer to first order in ε.

7.3 A quantum system is capable of existing in only two independent states ϕ_1 and ϕ_2, which are eigenstates of the Hamiltonian H_0 with eigenvalues E_1 and E_2, respectively. The system is modified by the addition of a term H_1 to the Hamiltonian. Find expressions for the new eigenvalues E'_1 and E'_2, (i) by exact calculation, (ii) in lowest-order perturbation theory. Verify that the perturbation-theory result is an approximation to the exact result.

7.4 A particle of mass m and charge e is contained within a cubical box of side a. Initially the particle is in the stationary state of energy $3\hbar^2\pi^2/2ma^2$. At time $t = -\infty$ a uniform electric field is switched on parallel to one of the edges of the cube. Obtain the expression in lowest-order perturbation theory for the probability of eventually finding the particle in a state of energy $3\hbar^2\pi^2/ma^2$, assuming that the electric field varies with time as $\mathscr{E}\,\mathrm{e}^{-t^2/T^2}$.

7.5 A charged particle is scattered by an electron bound to a hydrogen atom. The problem is considered in terms of a crude one-dimensional model in which the electron is described by a harmonic-oscillator wave function and the charged particle induces the perturbation $H_1 = \lambda\delta(x - vt)$. Show that the probability that the electron is excited from the ground state $n = 0$ to the $n = 1$ state is $\pi\lambda^2\hbar^6/2v^8m^8\omega^6$.

7.6 A particle is in the ground state in a one-dimensional box (given by the infinite square-well potential (2.11)). What is the probability that it remains in the ground state when the wall separation is suddenly halved?

8

Radiative transitions

When a system changes its energy as the result of the emission or the absorption of a photon, it is said to undergo radiative transition. There are three kinds of radiative transition. In the presence of an electromagnetic field the system can *absorb* a photon, so that its energy is raised to a higher level. If the system is initially in a state other than the ground state, it can emit a photon and so shed energy. This can happen without the presence of any external electromagnetic field, in which case a *spontaneous emission* is said to occur. On the other hand, if the excited system is placed in an electromagnetic field that varies in time with the appropriate frequency, the probability that it emits a photon can be greatly enhanced. In this case the emission process is known as *stimulated emission*. Stimulated emission is the basis of maser and laser action.

In this chapter we construct a simple model that allows us to investigate how the internal structure of a quantum system changes as the result of a radiative transition. Although the model is very simple, it will allow us to understand the main features of the proper, more exact treatment. We shall deal mainly with absorption and stimulated emission, the two types of radiative transition where there is an external electromagnetic field.

The electromagnetic interaction

In practice, the intensity of the external electromagnetic field is usually large compared with that of the field associated with the single photon that is being absorbed or emitted. This means that we may neglect the change in the total energy of the field that results from the absorption or emission. When the external field is so weak that this approximation is not valid, proper account must be taken of its quantum character. That is, it is then necessary to treat the external field as a collection of photons rather than as a classical continuous field. This is beyond the scope of this book; here we describe instead a *semiclassical* calculation, where the external field in not quantised.

We are interested in the interaction of the electromagnetic field with a charged particle, for example with an electron in an atom. This means that we must incorporate the effect of the field into the Hamiltonian for the quantum system that we are studying. It is found that the correct way to do this is to follow the same procedure as in classical physics.

In classical mechanics, a Hamiltonian H is used to give the equations of motion of a system by eliminating $\boldsymbol{p}$ in the equations

$$\frac{\mathrm{d}\boldsymbol{r}}{\mathrm{d}t}=\frac{\partial H}{\partial \boldsymbol{p}}, \qquad \frac{\mathrm{d}\boldsymbol{p}}{\mathrm{d}t}=-\frac{\partial H}{\partial \boldsymbol{r}}. \tag{8.1}$$

These equations are known as *Hamilton's equations*. For example, for a particle moving in a potential $V(\boldsymbol{r})$, where the Hamiltonian is

$$H_0=(p^2/2m)+V(\boldsymbol{r}), \tag{8.2}$$

Hamilton's equations give

$$\dot{\boldsymbol{r}}=\boldsymbol{p}/m, \qquad \dot{\boldsymbol{p}}=-\boldsymbol{\nabla} V,$$

and together these equations give, correctly, $m\ddot{\boldsymbol{r}}=-\boldsymbol{\nabla} V$. Suppose that the particle is an electron, so that it has charge $-e$, and that we switch on an external electromagnetic field. This field may be described by a scalar potential $\phi(\boldsymbol{r}, t)$ and a vector potential $\boldsymbol{A}(\boldsymbol{r}, t)$, so that its electric and magnetic vectors are given by

$$\boldsymbol{\mathscr{E}}=-\boldsymbol{\nabla}\phi-\dot{\boldsymbol{A}}, \qquad \boldsymbol{B}=\boldsymbol{\nabla}\wedge\boldsymbol{A}. \tag{8.3}$$

(See appendix D.) The Hamiltonian then changes from H_0 to

$$H=V+(\boldsymbol{p}+e\boldsymbol{A})^2/2m-e\phi. \tag{8.4}$$

In order to verify this, one shows that Hamilton's equations (8.1), together with (8.3), now give

$$m\ddot{\boldsymbol{r}}=-\boldsymbol{\nabla} V-e\boldsymbol{\mathscr{E}}-e\dot{\boldsymbol{r}}\wedge\boldsymbol{B}. \tag{8.5}$$

We do this in appendix D. In (8.5), the last two terms are just the correct Lorentz force for the action of the electromagnetic field on the charged electron.

In quantum mechanics, the Hamiltonian (8.4) becomes an operator, because $\boldsymbol{p}$ must be replaced by the operator $\hat{\boldsymbol{p}}=-\mathrm{i}\hbar\boldsymbol{\nabla}$. The vector potential $\boldsymbol{A}$ is not fully determined from the definitions (8.3), and it is convenient to impose on it the further *gauge condition* $\nabla \cdot \boldsymbol{A} = 0$ (see appendix D). This has the consequence that, when we multiply out the term $(\hat{\boldsymbol{p}} + e\boldsymbol{A})^2$ in (8.4), the order in which we write $\hat{\boldsymbol{p}}$ and $\boldsymbol{A}$ in their dot product does not matter. For if Ψ is any wave function,

$$(\hat{\boldsymbol{p}} \cdot \boldsymbol{A} - \boldsymbol{A} \cdot \hat{\boldsymbol{p}})\Psi = -\mathrm{i}\hbar\nabla \cdot (\boldsymbol{A}\Psi) + \mathrm{i}\hbar\boldsymbol{A} \cdot \nabla\Psi = -\mathrm{i}\hbar(\nabla \cdot \boldsymbol{A})\Psi = 0.$$

We can therefore write $(\hat{\boldsymbol{p}} + e\boldsymbol{A})^2 = \hat{\boldsymbol{p}}^2 + 2e\hat{\boldsymbol{p}} \cdot \boldsymbol{A} + e^2\boldsymbol{A}^2$. In most applications, it turns out that the term $e\hat{\boldsymbol{p}} \cdot \boldsymbol{A}$ is rather larger than $e^2\boldsymbol{A}^2$, which can therefore be neglected. So then (8.2) and (8.3) give

$$\begin{aligned} H &= H_0 + H_1 \\ H_1 &= (e/m)\hat{\boldsymbol{p}} \cdot \boldsymbol{A} - e\phi = (e/m)\boldsymbol{A} \cdot \hat{\boldsymbol{p}} - e\phi. \end{aligned} \tag{8.6}$$

Resonant absorption and stimulated emission

We consider the interaction between an electron bound to an atom and an electromagnetic wave. In the absence of the electromagnetic field, the electron wave function satisfies the Schrödinger equation with Hamiltonian H_0. Let the stationary-state wave functions of H_0 be $\Phi_k = \phi_k\, \mathrm{e}^{-\mathrm{i}E_k t/\hbar}$, corresponding to energies E_k. In reality, only the ground state is exactly a stationary state, because the higher states can spontaneously emit radiation and so change their configuration. By taking all the levels E_k to be stationary states, we are making an approximation in which the possibility of spontaneous emission is neglected. Our calculation will consider only the stimulated emission (and absorption) induced by the external electromagnetic field.

We shall take this field to be the plane wave given by

$$\phi = 0, \qquad \boldsymbol{A} = \boldsymbol{A}_0 \sin(\boldsymbol{K} \cdot \boldsymbol{r} - \omega t). \tag{8.7}$$

Then, according to (8.3), the electric and magnetic vectors are

$$\begin{aligned} \boldsymbol{\mathscr{E}} &= \omega\boldsymbol{A}_0 \cos(\boldsymbol{K} \cdot \boldsymbol{r} - \omega t) \\ \boldsymbol{B} &= \boldsymbol{K} \wedge \boldsymbol{A}_0 \cos(\boldsymbol{K} \cdot \boldsymbol{r} - \omega t). \end{aligned} \tag{8.8}$$

In order to satisfy Maxwell's equations (which are listed in appendix D), we must impose the constraints.

$$\omega = cK, \qquad \boldsymbol{K} \cdot \boldsymbol{A}_0 = 0. \tag{8.9}$$

The second of these equations also makes $\boldsymbol{A}$ satisfy the gauge condition $\nabla \cdot \boldsymbol{A} = 0$. We shall choose the directions of the coordinate axes such that the direction of propagation, parallel to the wave vector $\boldsymbol{K}$, is in the z-direction, and such that $\boldsymbol{A}_0$ is in the x-direction. Then, from (8.6) and (8.7),

$$\begin{aligned} H_1 &= (eA_0/m)\hat{p}_x \sin(Kz-\omega t) \\ &= (eA_0/2m\mathrm{i})\hat{p}_x\,[\mathrm{e}^{\mathrm{i}(Kz-\omega t)}-\mathrm{e}^{-\mathrm{i}(Kz-\omega t)}]. \end{aligned} \tag{8.10}$$

We suppose that the field is weak enough to allow us to use the time-dependent perturbation theory developed in the last chapter. Imagine that the field is switched on at time $t=0$ and is switched off again at some time T. We suppose that the electron was initially in the eigenstate Φ_i of H_0, and calculate the probability that a measurement, after the field has been switched off, will find that a transition has occurred to some other eigenstate of H_0. This transition probability (see (7.20) and (7.21)) is the squared modulus of

$$a_{if} = -\frac{\mathrm{i}}{\hbar}\int_0^T \mathrm{d}t\,\{u_{fi}^{(+)}\,\mathrm{e}^{\mathrm{i}(\omega_{fi}-\omega)t} - u_{fi}^{(-)}\,\mathrm{e}^{\mathrm{i}(\omega_{fi}+\omega)t}\}, \tag{8.11}$$

where

$$\begin{aligned} \omega_{fi} &= (E_f-E_i)/\hbar, \\ u_{fi}^{(\pm)} &= (eA_0/2m\mathrm{i})(\phi_f,\,\hat{p}_x\,\mathrm{e}^{\pm\mathrm{i}Kz}\,\phi_i). \end{aligned} \tag{8.12}$$

If $E_f > E_i$, the transition must have occurred through the absorption of energy from the electromagnetic field, and the more complete treatment in which the field is quantised shows that, in fact, the electron has absorbed a single photon from the field. If $E_f < E_i$, the electron has emitted a single photon. In either case, we see that the transition probability is proportional to the square of the intensity of the field. In the case of absorption, it is easy to understand intuitively that the probability should increase with the field intensity, but the result is perhaps surprising for the case of emission: the presence of the field stimulates the electron to emit a photon.

The integral (8.11) gives

$$a_{if} = \frac{1}{\hbar}\left[\frac{(1-\mathrm{e}^{\mathrm{i}(\omega_{fi}-\omega)T})u_{fi}^{(+)}}{\omega_{fi}-\omega} - \frac{(1-\mathrm{e}^{\mathrm{i}(\omega_{fi}+\omega)T})u_{fi}^{(-)}}{\omega_{fi}+\omega}\right]. \tag{8.13}$$

Suppose that the angular frequency ω of the external field is chosen such that $\hbar\omega$ is close to the difference between the energies E_i and E_f of the initial and final states. Then the denominator of one of the two terms in (8.13) is small, and so ω has been tuned to give a resonance in the transition probability. The term with the small denominator then dominates over the other term, which may be neglected in comparison with it. In the case of absorption, where $E_f > E_i$, so that $\omega_{fi} > 0$, the first term becomes dominant and the transition probability is

$$|a_{if}|^2 = \frac{4|u_{fi}^{(+)}|^2}{\hbar^2} \frac{\sin^2\frac{1}{2}(\omega_{fi}-\omega)T}{(\omega_{fi}-\omega)^2}. \tag{8.14}$$

In the case of emission, where $\omega_{fi} < 0$, the second term in (8.13) leads to an analogous expression.

We have already encountered a function very similar to (8.14): see (7.25) and figure 7.1. Unless ω is within a distance of about $2\pi/T$ from ω_{fi}, the transition probability is very small. In practice, ω_{fi} is often in the optical-frequency range, so that when T is equal to one second, the transition occurs readily only if ω differs from ω_{fi} by less than one part in 10^{14}. (However, this estimate ignores the fact that excited levels of atoms have a finite lifetime and so, because of the energy uncertainty principle which we described in chapter 7, their energy is not precisely determined. Allowing for this typically increases the frequency spread to about one part in 10^5.)

Electric dipole transitions

In order to evaluate the transition probability (8.14), it is necessary to calculate $u_{fi}^{(+)}$. Consider a hydrogen atom for definiteness, and take the origin at the nucleus of the atom. Then if ϕ_i and ϕ_f are electron wave functions corresponding to low-lying atomic levels, they are small unless $Kr \ll 1$ (see problem 8.1). Hence it is a good approximation to replace the exponential in the expression (8.12) for $u_{fi}^{(\pm)}$ by unity:

$$u_{fi}^{(\pm)} = (eA_0/2m\mathrm{i})(\phi_f, \hat{p}_x\phi_i). \tag{8.15}$$

Now, according to the commutation relations (4.28),

$$[x, \hat{p}_x] = \mathrm{i}\hbar, \qquad [x, \hat{p}_y] = [x, \hat{p}_z] = 0. \tag{8.16}$$

Pre-multiply and post-multiply the first of these relations in turn by $\hat{p}_x$; this gives two relations which when added together yield

$$[x, \hat{p}_x^2] = 2\mathrm{i}\hbar\hat{p}_x.$$

Similarly, the other two relations in (8.16) give

$$[x, \hat{p}_y^2] = [x, \hat{p}_z^2] = 0.$$

Since $H_0 = (\hat{p}_x^2 + \hat{p}_y^2 + \hat{p}_z^2)/2m + V(\boldsymbol{r})$, and $V(\boldsymbol{r})$ trivially commutes with x, we thus have

$$\hat{p}_x = (m/\mathrm{i}\hbar)[x, H_0] = (m/\mathrm{i}\hbar)(xH_0 - H_0x). \tag{8.17}$$

Inserting this relation into (8.15), and using the fact that ϕ_i and ϕ_f are eigenfunctions of H_0, we obtain

$$u_{fi}^{\pm} = [A_0(E_i - E_f)/2\hbar](\phi_f, -ex\phi_i). \tag{8.18}$$

Now, $-ex$ is the x-component of the electric dipole moment of the atom. For this reason, the transitions corresponding to the approximation we have made, that $e^{\pm iKz} \approx 1$, are known as *electric dipole* transitions. If we had not replaced the exponential in (8.12) by unity, we should have obtained additional terms corresponding to electric quadrupole and higher multipole transitions, and also terms corresponding to magnetic dipole and multipole moments. The magnetic terms may be understood from the fact that, crudely speaking, the electron in orbit round the nucleus is equivalent to an electric current loop, which gives the atom a magnetic moment that interacts with the electromagnetic field.

The transition probabilities are greatest for the electric dipole transitions. But for many pairs of levels E_i and E_f the quantity $(\phi_f, -ex\phi_i)$ vanishes: compare the result (7.24) for the case of a particle in a box. This leads to certain *selection rules*: radiative transitions readily occur only between certain pairs of atomic levels. It may be shown that spontaneous emission is subject to exactly the same selection rules. In consequence, the corresponding lines in the emission or absorption spectrum of the atom are much more pronounced than those that are associated with transitions that can occur only via a magnetic or higher electric multipole mechanism.

The transition process†

We now study in more detail the process that causes the electric dipole transition to occur.

Recall first that the results (8.13) and (8.14) for a_{if} obtained from perturbation theory are valid only so long as $|a_{if}| \ll 1$. If we choose ω to be exactly on resonance for absorption, $\omega = \omega_{fi}$, then (8.14) becomes proportional to T^2. So when T is larger than some value (which is inversely proportional to the field intensity A_0) the results obtained from perturbation theory are not valid.

In this circumstance, we must make a different type of approximation if we wish to calculate a_{if}. We return to the equations (7.18), which are exactly equivalent to the time-dependent Schrödinger equation. If

† The remainder of this chapter may be omitted at first reading.

there is no other energy level E_k close to E_f, the probabilities of transition to all the levels except E_f will remain small, and the perturbation-theory estimates of the corresponding coefficients a_{ik} will correctly predict them to be small. Hence, on the right-hand side of (7.18) we need retain in the summation over j only the terms $j=i$ and $j=f$. The equations (7.18), written for the two choices $k=i$ and $k=f$, then become

$$\begin{aligned} i\hbar\dot{a}_{ii}(t) &= (\Phi_i, H_1(t)\Phi_i)a_{ii} + (\Phi_i, H_1(t)\Phi_f)a_{if} \\ i\hbar\dot{a}_{if}(t) &= (\Phi_f, H_1(t)\Phi_i)a_{ii} + (\Phi_f, H_1(t)\Phi_f)a_{if}. \end{aligned} \tag{8.19}$$

We use the form (8.10) for H_1 and the definitions (8.12), and set $\omega = \omega_{fi}$. Further approximations enable us to simplify the equations (8.19) and to write them in the form (8.22) given below.

We write the two coupled differential equations that result from (8.19) in matrix form. With

$$\begin{pmatrix} a_{ii} \\ a_{if} \end{pmatrix} = \boldsymbol{a},$$

they are

$$i\hbar\dot{\boldsymbol{a}}(t) = \boldsymbol{M}(t)\boldsymbol{a}(t), \tag{8.20}$$

where the 2×2 matrix $\boldsymbol{M}$ has the form

$$\boldsymbol{M}(t) = \boldsymbol{M}_0 + \boldsymbol{M}_1\,e^{i\omega t} + \boldsymbol{M}_2\,e^{-i\omega t} + \boldsymbol{M}_3\,e^{2i\omega t} + \boldsymbol{M}_4\,e^{-2i\omega t}, \tag{8.21a}$$

with

$$\boldsymbol{M}_0 = \begin{pmatrix} 0 & u_{if}^{(-)} \\ u_{fi}^{(+)} & 0 \end{pmatrix}. \tag{8.21b}$$

The precise expressions for the other matrices in (8.21a) will not be needed; we merely note that each is independent of t.

There is a well-established procedure for solving a system of first-order coupled differential equations such as (8.20). However, we shall simplify the equations by retaining, in the expression (8.21a) for $\boldsymbol{M}$, the term $\boldsymbol{M}_0$ only. In order to understand why this is a valid approximation, consider the simpler problem of a single differential equation, $i\hbar\dot{a}(t) = M(t)a(t)$. This is solved by the standard integrating-factor procedure; the solution is $a(t) = a(0)\,e^{F(t)}$, where

$$F(t) = \frac{1}{i\hbar}\int_0^t d\tau\, M(\tau).$$

If $M(t) = M_0 + M_1\,e^{i\omega t}$, say, then

$$F(t) = (M_0 t/i\hbar) - (M_1/\hbar\omega)(e^{i\omega t} - 1).$$

For radiation in the range of visible frequencies, ω is very large, and in practice one is interested in values of t that satisfy $t \gg \omega^{-1}$. Then the first term in $F(t)$ becomes dominant and the high-frequency term $M_1\,e^{i\omega t}$ of M becomes unimportant. Similarly, we may neglect the high-frequency terms in the expression (8.21*a*) for the matrix $\boldsymbol{M}$.

With the neglect of high-frequency terms, the coupled equations (8.19) become

$$\begin{aligned} i\hbar\dot{a}_{ii}(t) &= -u_{if}^{(-)}\,a_{if} \\ i\hbar\dot{a}_{if}(t) &= u_{fi}^{(+)}\,a_{ii}. \end{aligned} \tag{8.22}$$

The quantity $u_{fi}^{(+)}$ is given in (8.18), and $u_{if}^{(-)}$ is obtained by interchanging the labels i and f in (8.18). We shall suppose that $u_{fi}^{(+)}$ is real and positive. If this is not so, it can always be made so by multiplying one of the wave functions, ϕ_f say, by a suitable constant phase factor $e^{i\delta}$ (this will not change the normalisation $(\phi_f, \phi_f) = 1$). Then $u_{if}^{(-)} = -u_{fi}^{(+)}$.

The solution to the equations (8.22) that satisfies the required boundary conditions $a_{ii}(0) = 1$ and $a_{if}(0) = 0$ (so that the wave function is initially equal to Φ_i) is then

$$a_{ii} = \cos \Omega t, \qquad a_{if} = -i \sin \Omega t,$$

where

$$\Omega = u_{fi}^{(+)}/\hbar. \tag{8.23}$$

According to (7.16), the wave function of the electron at time t is

$$\Psi(\boldsymbol{r}, t) = a_{ii}\Phi_i + a_{if}\Phi_f. \tag{8.24}$$

(Recall that the coefficients a_{ij} are small for $j \neq i$ or f, and that we are neglecting them.) So we see from (8.23) that the atom oscillates back and forth between the two configurations Φ_i and Φ_f. It begins at $t = 0$ in the state Φ_i, and by the time $t = \pi/2\Omega$ a measurement would be certain to find that it has made a transition to the state Φ_f. Then at time $t = \pi/\Omega$ the atom is back in the state Φ_i, and so on. The frequency of this oscillation is very much less than the frequency of the radiation. Each time that the transition is upwards in energy, a photon is absorbed from the electromagnetic field, while downward transitions are associated with the emission of a photon into the field.

From (8.23) and (8.24), the electron probability density at time t is

$$\begin{aligned} |\Psi(\boldsymbol{r}, t)|^2 = |\phi_i(\boldsymbol{r})|^2 \cos^2 \Omega t + |\phi_f(\boldsymbol{r})|^2 \sin^2 \Omega t \\ + \mathrm{Re}\,[i\phi_f^*(\boldsymbol{r})\phi_i(\boldsymbol{r})\,e^{-i\omega t}] \sin 2\Omega t. \end{aligned} \tag{8.25}$$

Here we have supposed that $E_f > E_i$, so that $\omega = (E_f - E_i)/\hbar$. Now, according to (8.23) and (8.18), Ω is proportional to the amplitude A_0 of the electromagnetic field. For realistic values of A_0, it is found that

$\Omega \ll \omega$. Hence, because of the exponential inside the square brackets, the last term oscillates through very many cycles in a timescale in which the other two terms hardly vary at all. At any instant, the electric dipole moment of the atom is $(\Psi, -ex\Psi)$, and the last term in (8.25) gives a contribution to this which varies with the frequency of the radiation, namely

$$\mathrm{Re}\,[\mathrm{i}(\phi_f, -ex\phi_i)\,\mathrm{e}^{-\mathrm{i}\omega t}]\sin 2\Omega t.$$

Because we have supposed that the phase of the wave function ϕ_f is such that $u_{fi}^{(+)}$, given by (8.18), is real and positive, this is

$$-|(\phi_f, -ex\phi_i)|\sin 2\Omega t \sin \omega t. \tag{8.26}$$

The physics must be unaffected by the phase chosen for the wave function, and the result (8.26) is actually independent of this phase.

We recall that in using the approximation (8.18) for $u_{fi}^{(\pm)}$, in which the exponentials $\mathrm{e}^{\pm\mathrm{i}Kz}$ in (8.12) are replaced by unity, we are treating the radiation in what is called the electric dipole approximation. We see from (8.26) that, as the atom emits and absorbs energy, its electric dipole moment varies with the same frequency ω as the radiation. While the atom is in process of transition upwards in energy from level E_i to level E_f, for example during $0 < t < \pi/2\Omega$, the factor $\sin 2\Omega t$ is positive; while the atom is in process of downward transition, for example during $\pi/2\Omega < t < \pi/\Omega$, this factor is negative. That is, the phase of the dipole moment relative to that of the field is different according to whether the oscillating dipole moment is causing the absorption of radiation or its emission.

Problems

8.1 The wave number K corresponds to the radiation associated with transitions between the two lowest levels of the hydrogen atom (see (5.17)). Verify that the ground-state wave function (5.18) is small for all values of r that do not satisfy $Kr \ll 1$.

8.2 The vector potential corresponding to a uniform magnetic field $\boldsymbol{B}$ may be taken as $\boldsymbol{A} = \frac{1}{2}\boldsymbol{r} \wedge \boldsymbol{B}$. Show that when a hydrogen atom is placed in a uniform field parallel to the z-direction, a perturbation

$$H_1 = (-e\hbar/2M)BL_z$$

is added to the Hamiltonian H_0.

Notice that the Hamiltonian $H = H_0 + H_1$ commutes with H_0, so that H_0 and H have common eigenstates. What are these eigenstates?

Deduce that the original degeneracy of the hydrogen-atom energy levels with respect to the magnetic quantum number is resolved by the magnetic field. (This effect is called the *Zeeman effect*.)

9

Masers and lasers

The ammonia molecule

The ammonia molecule is a bound state of an atom of nitrogen together with three hydrogen atoms (figure 9.1). Thus it is a very complicated system: there are four nuclei and, when the molecule has no net charge, ten electrons. A full calculation of the energy levels of this fourteen-body system is much too difficult to undertake, but the system has an important property which we can understand by making an analysis similar to that given in chapter 6, where we described the ionised hydrogen molecule.

Consider the ammonia molecule in a coordinate frame in which its centre of mass is at rest. To specify fully the states of motion of the fourteen constituent particles, a large number of parameters would be needed. But only one of these need concern us here, namely the total angular momentum. It turns out that in the ground state the total

Figure 9.1. Two configurations of the ammonia molecule (neither is a stationary state). $\boldsymbol{S}$ is the spin vector and $\boldsymbol{\mu}$ is the electric dipole moment.

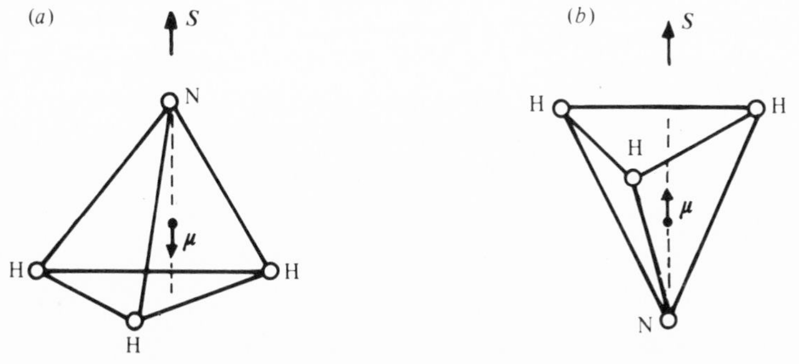

angular momentum of the molecule is non-zero. The total-angular-momentum vector is the sum of the orbital-angular-momentum and spin vectors of the separate constituent particles, and is called the *spin vector* of the molecule. In the ground state, the spin vector $\boldsymbol{S}$ is normal to the plane defined by the three hydrogen nuclei (that is, the expectation values of the spin components in directions parallel to the plane are zero). The direction in which $\boldsymbol{S}$ points distinguishes the two sides of this plane, so that the two configurations shown in figure 9.1, in which the nitrogen nucleus is positioned on opposite sides of the plane, correspond to different quantum-mechanical states. If the molecule had no spin, figure 9.1(*b*) would be obtained from figure 9.1(*a*) simply by turning the molecule over, and so would not correspond to a different internal state.

Let the wave functions ϕ_1 and ϕ_2 describe configurations of the molecule corresponding to figure 9.1(*a*) and figure 9.1(*b*), respectively. These wave functions differ only in that they make the expectation value of the coordinate of the nitrogen nucleus appear on different sides of the plane defined by the expectation values of the coordinates of the hydrogen nuclei. Because of the symmetry of the system,

$$H_{11}=H_{22} \quad \text{and} \quad H_{12}=H_{21}, \tag{9.1}$$

where $H_{ij}=(\phi_i, H\phi_j)$.

As in the discussion of the hydrogen molecule in chapter 6, the wave functions ϕ_1 and ϕ_2 do not describe stationary states. Their significance is merely that they represent configurations for which (9.1) applies. Because of the tunnel effect, transitions occur between these states, and as in chapter 6 the stationary states are

$$\psi_{\pm}=(\phi_1 \pm \phi_2)/\sqrt{2} \tag{9.2}$$

(see (6.3) and (6.6); the $1/\sqrt{2}$ is just a normalisation factor). The corresponding energies $E_{\pm}$ are to be calculated from a very complicated fourteen-particle Hamiltonian H and, unlike in chapter 6, the wave functions ϕ_1 and ϕ_2 are functions of many coordinates. The calculation is not possible in practice.

Without making a detailed calculation, it may however be shown that the ground state of the molecule corresponds to the wave function ψ_+. It is found experimentally that the level separation

$$E_- - E_+ = \hbar\omega_0 \tag{9.3}$$

corresponds to a frequency of 24 000 MHz. This is the frequency of the photons that are emitted or absorbed in a transition between the two levels. The corresponding wavelength, 12.5 mm, is in the microwave range. This level separation is very much less than the energy needed to excite the state of next-lowest energy. Thus, although the molecule has very many other stationary states, corresponding to, for example, rotational excitations of the whole molecule or excitations of its separate atoms, the molecule is unlikely to be found in one of these states unless the temperature is so high that there are frequent and violent collisions between molecules. If this were not the case, an analysis based on an expansion of the wave function similar to (6.3), in terms of ϕ_1 and ϕ_2 only, would not have been a good approximation.

The ammonia maser

The word *maser* is an acronym for *microwave amplification by stimulated emission of radiation*. The first maser was constructed in 1954, using ammonia gas.

At normal temperatures, very nearly equal numbers of molecules of the gas are found in each of the two lowest levels E_+ and E_-. The first stage in the operation of the maser is to separate, at least partially, those molecules in the lower level E_+ from those in the higher level E_-. The component enriched with molecules in the higher level is then squirted through a resonant cavity, tuned to angular frequency ω_0, so that some of the molecules emit photons of energy $\hbar\omega_0$ and emerge from the cavity in the lower level E_+. The emission process is stimulated emission, which we described in the last chapter, so that the number of photons emitted into the cavity by the gas increases if the intensity of the radiation already present in the cavity is increased. Hence the maser acts as an amplifier of any external radiation of frequency ω_0 that is fed into the cavity. (The maser can, to some extent, be tuned; that is, the frequency ω_0 at which it operates can be varied. To achieve this, one has to vary the energy gap $(E_- - E_+)$, for example by placing the maser in a magnetic field.)

We now consider in some detail the two stages in the operation of the maser.

The separation

The separation is achieved by passing the gas through a static electric field $\mathscr{E}$. This field interacts with the electric dipole moment $\boldsymbol{\mu}$ of the molecule. This dipole moment exists because the molecule is not spherically symmetrical. In fact, the electrons of the molecule tend to be found closer to the nitrogen nucleus than to the plane containing the hydrogen nuclei, so that the two states ϕ_1 and ϕ_2 have equal and opposite dipole moment, normal to the plane of the hydrogen nuclei, as shown in figure 9.1. Thus, if $\boldsymbol{\mu}_0$ is the expectation value of the dipole moment in the state ϕ_1,

$$\begin{aligned} H_{11} &= E_0 - \boldsymbol{\mu}_0 \cdot \boldsymbol{\mathscr{E}} \\ H_{22} &= E_0 + \boldsymbol{\mu}_0 \cdot \boldsymbol{\mathscr{E}}, \end{aligned} \tag{9.4}$$

where E_0 is the value of H_{11} or H_{22} in the absence of the electric field. We shall suppose that $\mathscr{E}$ is small compared with the internal field of the molecule; it would need a very strong external field $\boldsymbol{\mathscr{E}}$ for this not to be the case. This means that small changes in the geometry of the molecule due to the applied field $\boldsymbol{\mathscr{E}}$ can be neglected. Hence not only the dipole moment $\boldsymbol{\mu}_0$ can be taken to be unchanged by $\boldsymbol{\mathscr{E}}$, but also H_{12} and H_{21}, which, as in (6.13), depend mainly on the overlap $K_{12} = (\phi_1, \phi_2)$ between the wave functions ϕ_1 and ϕ_2.

In practice, $K_{12} \ll 1$, so that in the expressions (6.6) for $E_\pm$ we may neglect K_{12} compared with 1. According to (9.3), we define the difference between E_+ and E_- to be $\hbar\omega_0$, so that (6.6) gives

$$H_{12} = -\tfrac{1}{2}\hbar\omega_0. \tag{9.5}$$

If we now use the values (9.4) for H_{11} and H_{22} in the equations (6.4), eliminating c_1 and c_2 from them we find that in the presence of the electric field $\boldsymbol{\mathscr{E}}$ the energies $E_\pm$ are changed to

$$\begin{aligned} E'_\pm &= E_0 \mp [\tfrac{1}{4}\hbar^2\omega_0^2 + (\boldsymbol{\mu}_0 \cdot \boldsymbol{\mathscr{E}})^2]^{1/2} \\ &\approx E_0 \mp [\tfrac{1}{2}\hbar\omega_0 + (\boldsymbol{\mu}_0 \cdot \boldsymbol{\mathscr{E}})^2/\hbar\omega_0]. \end{aligned} \tag{9.6}$$

Thus, whatever the direction of $\boldsymbol{\mathscr{E}}$ relative to that of $\boldsymbol{\mu}_0$, the energy of the upper level E_- has been increased a little by the presence of the electric field, and the energy of the lower level E_+ has been decreased. The stationary-state wave functions also will be changed a little from the $\psi_\pm$ of (9.2); however, since in practice $\mathscr{E}$ is small on the atomic scale, this change is small.

Suppose now that the magnitude of $\boldsymbol{\mathscr{E}}$ varies with position. In practice, $\mathscr{E}$ will still be constant on the atomic scale of length, so that we need not take account of its variation in the quantum-mechanical

Hamiltonian, and it is sufficient to account for it classically. That is, the two components of the gas will experience forces

$$-\nabla E'_{\pm} = \pm(1/\hbar\omega_0)\nabla(\boldsymbol{\mu}_0 \cdot \boldsymbol{\mathscr{E}})^2 \tag{9.7}$$

in opposite directions. So if the gas is squirted through a region where $\boldsymbol{\mathscr{E}}$ varies across the jet, the two components are deflected in opposite directions (figure 9.2). The component that contains the molecules with wave function ψ_-, in the upper of the two energy levels, is made to pass into the resonant cavity.

Stimulated emission in a radiation cavity

Consider two parallel conducting planes, $z = 0$ and $z = a$, with empty space between. A solution to Maxwell's equations (listed in appendix D) in the space between the planes is

$$\begin{aligned} \boldsymbol{\mathscr{E}} &= (\mathscr{E}_0 \sin Kz \cos \omega t, \quad 0, \quad 0) \\ \boldsymbol{B} &= (0, \quad (\mathscr{E}_0/c) \cos Kz \sin \omega t, \quad 0) \end{aligned} \tag{9.8}$$

where

$$\omega = cK.$$

Because the planes $z = 0$ and $z = a$ are conductors, we must have $\mathscr{E}_x = \mathscr{E}_y = 0$ on them. This requires that

$$K = n\pi/a, \qquad n \text{ integer.} \tag{9.9}$$

The need to satisfy the boundary conditions at $z = 0$ and $z = a$ has led us to consider a standing wave (9.8), instead of a travelling wave as in chapter 8. However, the analysis of the interaction of the radiation (9.8) with the ammonia molecule can be carried through in a way closely similar to that described in the last chapter.

Figure 9.2. Schematic diagram of the ammonia maser.

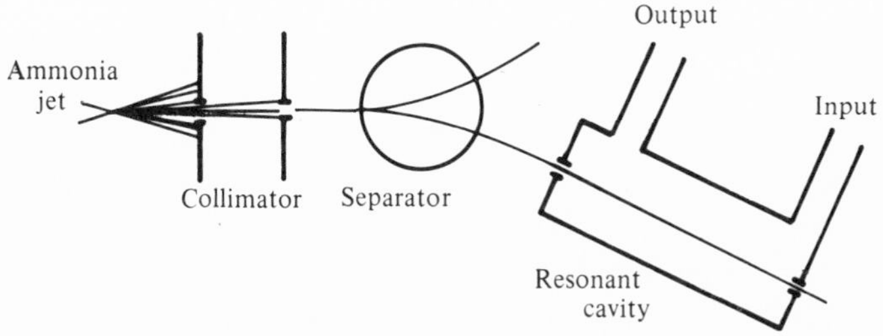

Suppose that a jet of ammonia gas enters the cavity with each of the molecules initially in the ψ_- state. The presence of an electromagnetic field (9.8) will stimulate the molecules to emit photons and so make transitions downwards in energy into the ψ_+ state. However, unless ω is tuned so as to be very close to ω_0, the probability of such transitions occurring will be very small. Because of (9.9) and the relation $\omega = cK$, this tuning may be achieved by suitable choice of the distance a between the conducting walls of the cavity. When ω is tuned in this way, the emission occurs fairly readily, and the photon that is emitted adds to the intensity of the electromagnetic field (9.8), so that the probability of further molecules undergoing stimulated emission is then enhanced. Each transition further reinforces the field and so increases the probability of yet more transitions occurring, so that the field builds up rapidly. That is, the original electromagnetic field (9.8) rapidly becomes amplified.

The standing wave (9.8) corresponds to a collection of photons that bounce back and forth between the two conducting planes. The momentum vector of each photon is normal to the planes, and when a photon hits one of the planes its momentum is reversed. What our simple discussion does not show is that each extra photon that results from the stimulated emission actually has the same wave vector, and therefore from (2.2) the same momentum, as the other photons that are already present in the cavity. It therefore exactly reinforces the existing field. To show this, one must consider in more detail the equivalence between electromagnetic fields and systems of photons, which is beyond the scope of this book.

The speed with which the ammonia gas passes through the cavity determines the time that the molecules spend in the resonant system. This is an important factor in the design of the maser. The reason for this is that once a molecule has undergone stimulated emission and so made a downward transition to the ground state ψ_+, until it has left the cavity it is now liable to *absorb* radiation and so make a transition back to the state ψ_-. If this is allowed to occur, it decreases the intensity of the field in the cavity instead of amplifying it. This problem is avoided by ensuring that the molecules which have dropped into the ground state escape through the exit hole before an appreciable probability of reabsorption occurs. The length of time that they spend in the cavity is chosen by making a calculation similar to that in the last part of chapter 8.

Population inversion

Because a molecule in the ground state ψ_+ is liable to make a transition that absorbs energy from the field instead of adding to its intensity, the initial separation of the gas into two components is a vital stage in the operation of the maser. If an unseparated, equal mixture of the two components ψ_+ and ψ_- enters the cavity, the net effect is zero: there are as many upward transitions that absorb photons as there are downward transitions that emit them.

Even a small quantity of gas contains a very large number of molecules – about 10^{25} kg^{-1}. The molecules continually collide with one another and so exchange energy. It is completely impracticable to try and follow the effect of the collisions on the state of each individual molecule. Instead, one resorts to statistical methods, and it turns out that this is quite sufficient to give an accurate description of the macroscopic properties of the gas. There is little interest in determining at any instant the state of a given molecule; what matters is the configuration of the system as a whole. That is, one needs to know *how many* molecules at any instant may be expected to be in a given quantum-mechanical state, rather than *which* particular molecules are in that state.

The molecules continually collide with each other and so change their states. The collisions become more frequent and more violent as the temperature of the gas increases, since by definition the average kinetic energy of the molecules is proportional to the temperature (measured on the absolute, or Kelvin, scale). A remarkable result of statistical physics is that it predicts the average distribution of the states of the molecules without knowing anything in detail about what happens in the collisions. The demonstration of this may be found in any book on statistical physics;† here we just quote the results.

Consider any system of particles that is in dynamical equilibrium, so that in particular no heat flows into or out of the system. Suppose that each particle may be found in any one of a number of different states. Let these have energies E_i, $i = 1, 2, \ldots$, with corresponding degeneracies g_i (so that the level E_i is associated with g_i independent states). Then according to classical statistical mechanics, the average

† For example, C. Kittel, *Thermal Physics* (Wiley, 1969).

number n_i of particles expected to be in the ith level at any given instant is given by the *Maxwell–Boltzmann distribution*:

$$n_i = C g_i \, e^{-E_i/k_B T}. \tag{9.10}$$

Here T is the absolute temperature and k_B is a universal constant, known as Boltzmann's constant. The constant C may be calculated in terms of the total number $N = \sum_i n_i$ of particles:

$$C = N \left(\sum_i g_i \, e^{-E_i/k_B T} \right)^{-1}. \tag{9.11}$$

Evidently C varies with the temperature T.

The Maxwell–Boltzmann distribution is a formula of classical mechanics. In certain circumstances it is important to take account of quantum-mechanical effects, but at room temperatures (9.10) adequately describes the distribution of states of the molecules in the gas. Because the energy difference between the levels E_+ and E_- of ammonia is small, we deduce from (9.10) that in a quantity of gas that is in dynamical equilibrium the numbers of molecules in these two levels are very nearly equal, the population of the lower level E_+ being only slightly larger than that of the higher level E_-. The separation process achieves a *population inversion* in the component of the gas that is injected into the cavity. In this component the majority of the molecules are in the higher level, and the gas is not in dynamical equilibrium. If it is left to return to equilibrium, intermolecular collisions ensure a rapid return to the Maxwell–Boltzmann distribution.

From (9.10) we may see also the truth of our earlier assertion that, unless the temperature T is high, the populations of the higher-energy levels are rather smaller than those of the lowest-lying levels E_+ and E_-.

The laser

The principles underlying the operation of the laser are much the same as those of the maser, the basic difference being that the photons produced correspond to the optical range of frequencies. In lasers, and nowadays also in masers, the material that emits the photons is kept stationary within the radiation cavity, instead of being squirted through the cavity. If the amplifying material is solid, the parallel mirrors that form the radiation cavity may be the end faces of the material itself. A resonant cavity of this form was used in the first successful laser, whose operation was announced in 1960. This laser had as its active medium a cylinder of synthetic ruby, which is an aluminium oxide crystal containing a small percentage of chromium. The end faces of the ruby cylinder were rendered accurately parallel to each other and then both were silvered, one fully and the other partially. Population inversion was achieved by a process known as *optical pumping*, which uses a helical flashtube wrapped around the ruby cylinder. The flashtube emits white light, that is, light with a wide range of frequencies.

The active ingredient of the ruby is the chromium, whose (simplified) energy-level diagram is shown in figure 9.3. When the flashtube is fired, the ruby crystal is flooded with photons, most of which dissipate their energy as heat. However, some in the blue and green ranges of the spectrum are absorbed by the chromium and cause transitions from the ground state to one of the two energy bands shown in the figure. We explained in chapter 8 that the relative probabilities of transitions between different levels of a system are governed by selection rules. In the case of the energy bands it turns out that the preferred decay is to a metastable state, also indicated in figure 9.3. (This decay is a non-radiative transition: the energy is not given up by the emission of a photon, but rather it appears in the form of *phonons*, which are quanta of vibration of the other atoms in the crystal lattice about their mean positions.) So if the flashtube is fired with high power, a large number of chromium ions are driven into the metastable state: a large population inversion occurs.

Figure 9.3. Energy levels in a ruby crystal.

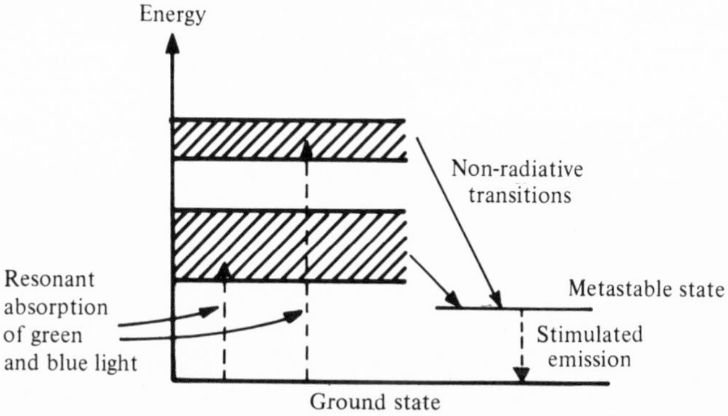

The metastable state decays to the ground state by spontaneous emission of photons. (Its lifetime is rather long on the atomic scale: a few milliseconds, compared with the 10^{-8} s typical of spontaneous transitions when dipole transitions are allowed.) Most of the photons emitted by spontaneous emission pass out of the crystal and are lost; but just a few move in the right direction to be reflected back and forth between the two silvered end faces. In this way, a standing-wave electromagnetic field of the form (9.8) begins to be set up. This field *stimulates* the emission of further photons, so that the intensity of the field (9.8) builds up very rapidly as the metastable state depopulates. The partially silvered end of the crystal allows the leakage of an intense pulse of laser light. This is the output of the pulsed ruby laser.

Laser technology has developed very rapidly since 1960, and there are now many types of laser.

The light emitted by a laser has two important properties. First, it can be very intense: the power per unit area of cross-section in a laser beam can be at least 10^9 times that which may be obtained from a conventional light source. Secondly, laser light is not only highly monochromatic, but also highly coherent. In a conventional light source, the atoms decay in an uncorrelated fashion and so the phases of the electromagnetic wave packets associated with the different photons are random. In a laser, the stimulated emission is precisely in phase with the electromagnetic field that causes it. There are some variations in phase of the field as a whole, mainly caused by fluctuations in the distance between the mirrors resulting from thermal and mechanical disturbances, but these are extremely small. In a laser beam, the phase difference between the light at points along the beam separated by as much as 100 km is likely to be unchanging with time.

Holography

The high intensity of laser light has obvious applications, such as to cutting and welding in engineering or medicine. It also makes possible the study of fundamental phenomena, such as non-linear optics, arising from the changes in the properties of materials when they are subjected to very strong electromagnetic fields. An important application of the high degree of coherence of laser light is to holography. We digress briefly from our main subject of quantum physics in order to describe this technique of three-dimensional photography.

Imagine first a coherent, monochromatic point source of light S at a distance R from a point object O (figure 9.4). Light that comes from S in the direction of O will be scattered by O, and hence O behaves effectively as a second point source. The two sources S and O will be mutually coherent; that is, the phase difference between the light from them will be constant in time at any point P in the space surrounding them. Hence, an interference pattern is established: if the two components of the light reaching P are exactly out of phase, then there is darkness at P, while if they are in phase, then P is a point of maximum brightness. The condition that there be destructive interference at P, that is, darkness, is

$$K(r_1-R-r_2)+\alpha=(2m+1)\pi \qquad (9.12a)$$

where

$$m=0,\ \pm 1,\ \pm 2,\ldots$$

Here K is the wave number of the light (so that its wavelength is $2\pi/K$), r_1 and r_2 are the distances of P from S and O as in figure 9.4, and α is a possible phase change induced by the reflection of the light from S on the object O. Similarly, the condition that there be constructive interference at P is

$$K(r_1-R-r_2)+\alpha=2m'\pi \qquad (9.12b)$$

where

$$m'=0,\ \pm 1,\ \pm 2,\ldots$$

The points P that satisfy either of the two conditions (9.12) lie on a system of hyperboloids of revolution having O and S as foci, as is shown in figure 9.5(a). In this figure we show also the cross-section of a photographic plate that has been placed so as to intersect the interference pattern. This plate records a *hologram*, which is a thin section of the intereference pattern. Development of the plate causes silver grains to be deposited in the emulsion, with a high density at points P corresponding to constructive interference, and low density in between at points of destructive interference.

To view the holographic image, one illuminates the hologram with a source S', having the same position relative to it as the original source S and emitting light of the same colour. Unlike S, the source S' does not have to be coherent. The light from S' that falls on the hologram is diffracted by the light and dark fringes recorded on it. Diffraction occurs in two directions, one such that the rays emerging on the other side of the hologram appear to have originated from the point O' corresponding to the position of the original object O, and the other such that the rays converge on the point O'' that is the reflection of O' in the photographic plate (see figure 9.5*b*). Thus there is a virtual

Figure 9.4. Light reaching a point P directly from a coherent source S and reflected from a point object O.

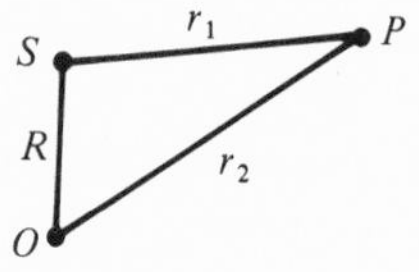

Figure 9.5. (*a*) Interference fringes corresponding to figure 9.4, with a photographic plate on which a hologram is recorded. (*b*) Reconstruction of a virtual image O' and a real image O'', using a light source S'.

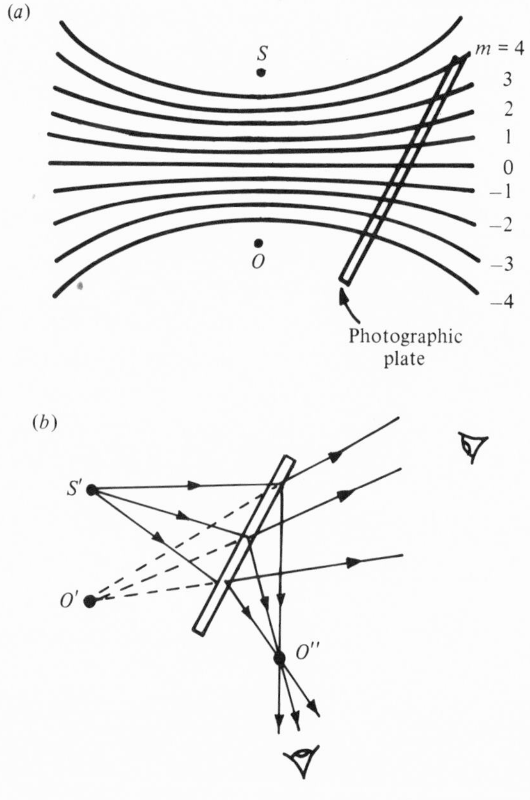

image at O' and a real image at O''. To view the virtual image, one looks at the plate from behind but focuses the eye on O' because this is the point from which the rays appear to have diverged; this is why holography produces a three-dimensional effect. The real image is made apparent by placing a screen behind the plate, passing through the point O'', or by placing the eye beyond the point O'' and focussing it on that point (see figure 9.5*b*).

In practice, because a suitably coherent point source of light *S* is not, in fact, available, a laser beam is used to produce the hologram. The beam is spread out by a lens and split into two parts. One part is reflected directly onto the photographic plate, and the other is reflected onto the plate by the object. The set-up needed to produce a hologram of an extended three-dimensional object is shown in figure 9.6. The interference fringes now have a much more complicated geometry, but the principles involved are the same as for a point object.

Figure 9.6. Set-up used to produce a hologram of a three-dimensional object.

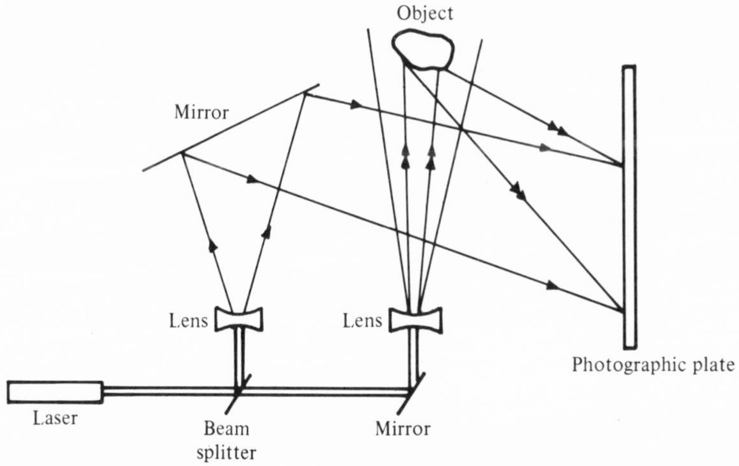

Problems

9.1 Assuming that the states ϕ_1 and ϕ_2 of the ammonia molecule are eigenstates of the electric-dipole-moment operator, show that the stationary states $\psi_\pm$ have zero dipole moment.

9.2 For a black body, the power emitted per unit area in the wavelength interval from λ to $\lambda + d\lambda$ is

$$(2\pi hc^2/\lambda^5)[d\lambda/(e^{hc/\lambda k_B T} - 1)].$$

A helium–neon laser operating at a wavelength of 633 nm with an output power of 1 mW has a linewidth of 1 kHz and produces a beam of cross-sectional area 10^{-6} m^2. What would be the temperature of the black body that emitted photons at the same rate, in the same frequency interval and from the same area?

9.3 The beam from a laser is usually extremely well collimated. By applying the uncertainty principle to estimate the transverse momentum of the photons, show that a helium–neon laser beam which initially has diameter 1 mm can be collimated to within an angle of about 3×10^{-3} degrees.

How small would the filament of an ordinary household torch have to be to achieve the same collimation?

9.4 A gas of non-interacting particles of mass m is contained in a cube of side L and obeys the Maxwell–Boltzmann distribution. Write down the formula for the probability that a given particle is in the state corresponding to wave function $(8/L^3)^{1/2} \sin q\pi x/L \sin r\pi y/L \sin s\pi z/L$. Show that as $L \to \infty$ the probability that the x-component of the velocity of a given particle lies between u and $u + du$ is

$$du(m/2\pi k_B T)^{1/2} e^{-mu^2/2k_B T}.$$

9.5 In the previous question, a membrane of area A is embedded in one wall of the cube. It is permeable to all particles of the gas whose normal component of velocity is greater than u_0. Show that the rate at which particles escape from the box is

$$(NA/L^3)(k_B T/2\pi m)^{1/2} e^{-mu_0^2/2k_B T}.$$

9.6 In quantum statistical mechanics for a system of electrons, the Maxwell–Boltzmann distribution is replaced by the *Fermi–Dirac distribution*

$$n_i = g_i/[e^{(E_i - \zeta)/k_B T} + 1],$$

where the *Fermi energy* ζ is independent of E_i. What form does this distribution take at $T = 0$, and why?

Notice that when $E_i \gg \zeta$ the two distributions are approximately the same.

10

Band structure of crystals

Electrons in crystals

A crystal consists of a collection of atoms arranged in a regular array, the spacing between atoms being of the same order of magnitude as the dimensions of the atoms. Each atom is more or less anchored to one point, called its *site* in the lattice, by the electrostatic forces produced by all the other atoms. We shall not find it necessary here to discuss the details of how this comes about; nor shall we consider the various patterns in which the atoms can be arranged. It will be sufficient to remember the essential feature that the structure of the crystal is periodic in space.

We have seen in chapter 5 that the energy of an electron bound to an atom is restricted to certain discrete values. Imagine that we can assemble a crystal of identical atoms whose spacing L can be altered at will. If L is large enough, the motion of an electron in one of the atoms will be affected to a negligible extent by the electrons and nuclei of the other atoms. Each atom then behaves as if it were isolated, with its electrons in discrete bound states. In figure 10.1(a) we have drawn a schematic diagram of the potential $V(\boldsymbol{r})$ in which an electron moves in this situation. Suppose that the spacing L is now reduced (figure 10.1(b)). The potential $V(\boldsymbol{r})$ in the neighbourhood of a given atom is now affected by the presence of the nuclei and electrons of the other atoms, particularly those that are closest. An exact calculation of this situation is a complicated many-body problem and is not tractable. To a reasonable approximation, however, the motion of a given individual electron in a crystal can be treated as a single-particle problem, with the potential determined by the electric field produced by all the other particles in the crystal.

We shall show in this chapter that, as L is reduced, each original discrete atomic energy level spreads out into a *band* of closely spaced levels (figure 10.2). These bands are separated by *energy gaps* that are forbidden to the electrons. The band structure is all-important in determining the properties of electrons in crystals.

Figure 10.1. Schematic diagram of the potential in a crystal (*a*) when the atoms are widely separated; (*b*) when they are closer together.

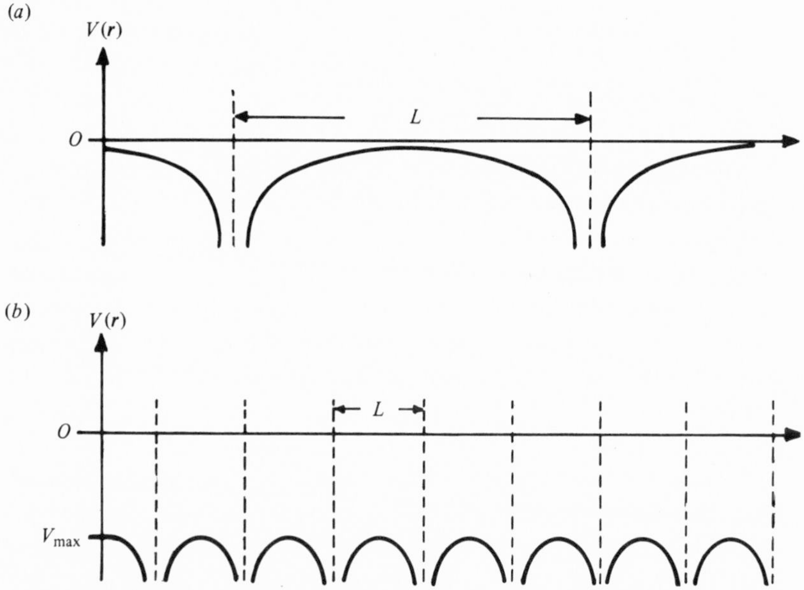

Figure 10.2. The band structure that develops when the atomic separation L is decreased.

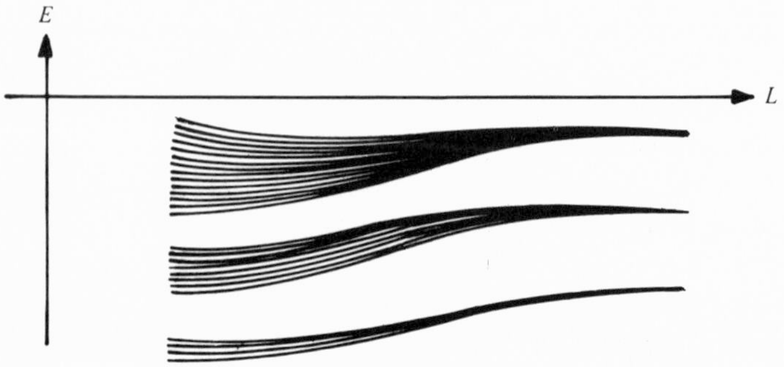

In a crystal, an electron is not bound to any one particular atom, but rather to the crystal as a whole. If it is not to escape from the crystal, its energy must be negative (where the zero of energy is defined to be the potential of the free space outside the crystal). If the crystal temperature is high enough, an electron can gain sufficient energy by collisions with the vibrating nuclei to escape from the crystal; this is called *thermionic emission.* But we shall confine the discussion to the electrons that remain bound in the crystal. An electron in one of the low-lying bands, such that its energy is less than V_{max} (see figure 10.1(b)), passes from one atom to another by tunnelling through the potential barriers between the nuclei. We shall find that there are also energy bands between V_{max} and 0; electrons in these bands are readily transmitted over the potential barriers.

To explore the band structure, we consider mainly a one-dimensional model. This provides a good guide to electron behaviour in real crystals. Certain aspects, however, require a discussion of electron motion in more than one dimension; this is necessary, for example, in order to understand why divalent metals such as calcium and magnesium are reasonably good electrical conductors.

Band structure

We do not need to know in detail how the potential varies with position in the crystal. The most important feature is its periodicity. To have a readily solvable problem, we consider an electron in a one-dimensional periodic square-wave potential (figure 10.3). This model is sufficiently realistic for many purposes. For convenience, we define the zero of the potential to coincide with the top of the wells, rather than with the value of the potential in free space outside the crystal.

Figure 10.3. One-dimensional periodic square-well potential.

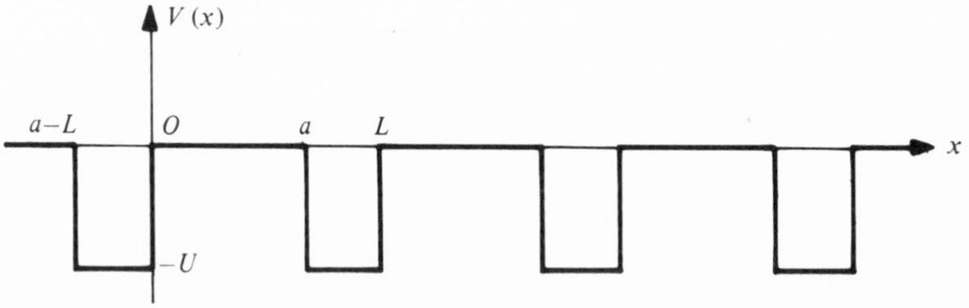

Consider first an electron with energy E such that $-U<E<0$. The Schrödinger equation is

$$\begin{aligned}(-\hbar^2/2m)\psi''-U\psi&=E\psi \quad \text{in the wells}\\ (-\hbar^2/2m)\psi''&=E\psi \quad \text{between the wells,}\end{aligned} \tag{10.1}$$

or

$$\begin{aligned}\psi''&=\alpha^2\psi \quad \text{between the wells}\\ \psi''&=-\beta^2\psi \quad \text{in the wells,}\end{aligned} \tag{10.2}$$

where

$$\alpha^2=-2mE/\hbar^2, \qquad \beta^2=(2m/\hbar^2)(E+U).$$

The procedure for solving (10.2) is suggested by a theorem due to Bloch, which is derived in appendix E. One looks for solutions of the form

$$\psi(x)=\mathrm{e}^{\mathrm{i}kx}u_k(x), \tag{10.3}$$

where k is real and $u_k(x)$ is periodic, with the same period L as the period of the crystal lattice. Bloch's theorem states that solutions of the type (10.3) form a complete set, at least in the limit where the total length of the crystal is taken to be infinite. Evidently (10.3) represents a wave of wavelength $2\pi/k$, whose amplitude $u_k(x)$ varies across each atomic site, but is the same from one atom to the next. The electron belongs not to one atom, but has equal chance of being found in the vicinity of any of them.

The solution to each of the two differential equations (10.2) consists of simple exponentials. We write

$$\begin{aligned}\psi&=A\,\mathrm{e}^{\mathrm{i}\beta x}+B\,\mathrm{e}^{-\mathrm{i}\beta x} \qquad a-L<x<0\\ \psi&=C\,\mathrm{e}^{\alpha x}+D\,\mathrm{e}^{-\alpha x} \qquad 0<x<a.\end{aligned} \tag{10.4}$$

Then from (10.3)

$$\begin{aligned}u_k&=A\,\mathrm{e}^{\mathrm{i}(\beta-k)x}+B\,\mathrm{e}^{-\mathrm{i}(\beta+k)x} \qquad a-L<x<0\\ u_k&=C\,\mathrm{e}^{(\alpha-\mathrm{i}k)x}+D\,\mathrm{e}^{-(\alpha+\mathrm{i}k)x} \qquad 0<x<a,\end{aligned} \tag{10.5}$$

and because we require $u_k(x)$ to be periodic, (10.5) determines $u_k(x)$, and so also $\psi(x)$, everywhere. In particular, for $a<x<L$ we replace x by $(x-L)$ in the first equation of (10.5):

$$u_k=A\,\mathrm{e}^{\mathrm{i}(\beta-k)(x-L)}+B\,\mathrm{e}^{-\mathrm{i}(\beta+k)(x-L)}. \tag{10.6}$$

At both edges of each well we require ψ and ψ' to be continuous, or equivalently u and u' to be continuous. The periodicity of $u_k(x)$ guarantees that if these conditions are satisfied at the edges of just one well, then they are satisfied for every well. This, of course, is the reason for looking for a solution of the Bloch type (10.3). Imposing the conditions at $x=0$ on the two functions in (10.5) we have

$$\begin{aligned} A+B&=C+D \\ \mathrm{i}(\beta-k)A-\mathrm{i}(\beta+k)B&=(\alpha-\mathrm{i}k)C-(\alpha+\mathrm{i}k)D. \end{aligned} \tag{10.7}$$

From the conditions applied at $x=a$ to the second function in (10.5) and to (10.6),

$$A\,\mathrm{e}^{-\mathrm{i}(\beta-k)b}+B\,\mathrm{e}^{\mathrm{i}(\beta+k)b}=C\,\mathrm{e}^{(\alpha-\mathrm{i}k)a}+D\,\mathrm{e}^{-(\alpha+\mathrm{i}k)a}$$

$$\begin{aligned} &\mathrm{i}(\beta-k)A\,\mathrm{e}^{-\mathrm{i}(\beta-k)b}-\mathrm{i}(\beta+k)B\,\mathrm{e}^{\mathrm{i}(\beta+k)b} \\ &\quad=(\alpha-\mathrm{i}k)C\,\mathrm{e}^{(\alpha-\mathrm{i}k)a}-(\alpha+\mathrm{i}k)D\,\mathrm{e}^{(\alpha+\mathrm{i}k)a}, \end{aligned} \tag{10.8}$$

where

$$b=L-a. \tag{10.9}$$

The equations (10.7) and (10.8) are four homogeneous equations in four unknowns, A, B, C and D. In order that they have a nontrivial solution the determinant of their coefficients must vanish. When this 4×4 determinant is expanded it results in the condition

$$[(\alpha^2-\beta^2)/2\alpha\beta]\sinh\alpha a\sin\beta b+\cosh\alpha a\cos\beta b=\cos kL. \tag{10.10}$$

Although we have derived this equation for the case where E lies in the range $-U<E<0$, it is valid also when $E>0$; then α^2, defined in (10.2), is negative and in (10.10) we write $\alpha=\mathrm{i}\alpha'$ and so obtain

$$[-(\alpha'^2+\beta^2)/2\alpha'\beta]\sin\alpha' a\sin\beta b+\cos\alpha' a\cos\beta b=\cos kL. \tag{10.11}$$

According to (10.2), when a, L and U are fixed, the quantities α (or α') and β are functions of E, and so the left-hand sides of (10.10) and (10.11) correspond to a complicated function of E. We write these equations together as

$$f(E)=\cos kL \tag{10.12}$$

and in figure 10.4 we plot the function $f(E)$ against E for a typical set of values of a, L and U. Because k is real, $|\cos kL|\leq 1$. Thus those values of E for which $|f(E)|>1$ are not accessible; that is, the allowed values of E fall into bands determined by the condition $|f(E)|\leq 1$, with forbidden energy gaps between the bands.

Notice that, according to figure 10.4, the bands corresponding to the lowest-lying levels are narrowest: the levels associated with the electrons that are most tightly bound to the atoms are affected least by the presence of the other atoms. Notice also that the band structure persists even for positive values of E, that is, above the top of the potential wells. The energy gaps decrease in width with increasing E, but they can still be of appreciable width even when E is so large that the electron nearly has enough energy to escape from the crystal into the surrounding free space. As we shall describe below, when the level structure is calculated in more than one dimension, an alternative possibility is that neighbouring energy bands overlap, and this is found to be the case in certain types of crystal.

Number of levels in a band

Except for the smallest microcrystals, the properties of the crystal and the disposition of the allowed energy levels do not depend critically on the boundary conditions applied at its surfaces. The simplest choice of boundary condition is to require that the electrons be strictly confined within the crystal, that is $\psi = 0$ on the faces, or in the linear model $\psi = 0$ at each end of the chain of atoms. However, this has two disadvantages. First, the boundary conditions can then not be satisfied by

Figure 10.4. Plot of the function $f(E)$ defined in (10.12).

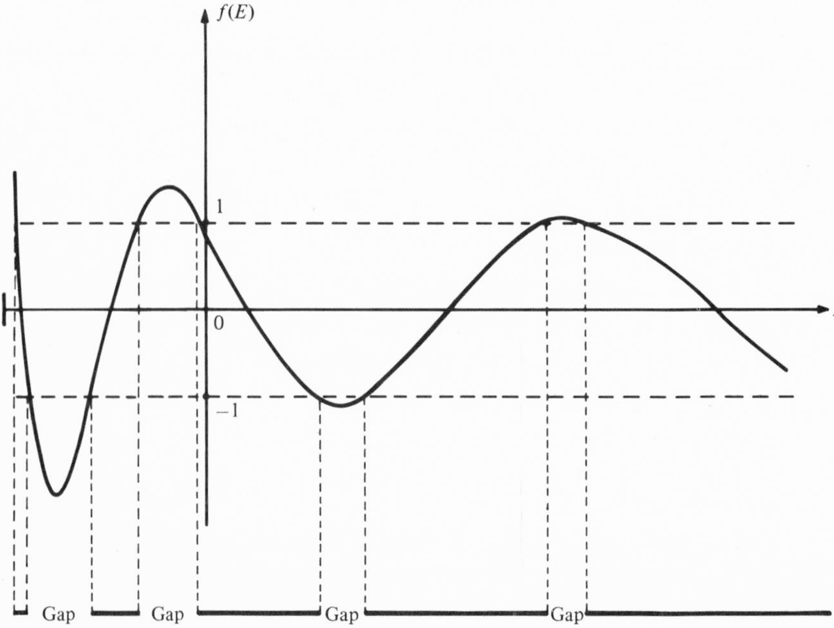

the simple Bloch wave functions (10.3); one must instead take combinations of Bloch waves to achieve a standing wave (a travelling wave of the type (10.3) moving in the positive-x direction is reflected at the crystal face, so generating another travelling wave moving in the negative-x direction). Secondly, a standing-wave solution is often unsatisfactory because transport of energy or charge is far more conveniently expressed in terms of travelling waves.

Since the crystal surface has little effect on the bulk properties of the crystal, one may as well choose boundary conditions that more or less eliminate it. This is achieved with *periodic boundary conditions.* The wave function ψ is required to take the same value at corresponding points on opposite faces of the crystal, so that an electron is not reflected at the faces but leaves the crystal and simultaneously re-enters it at the corresponding point on the opposite face. Both types of boundary condition result in the same number of levels within each band, but they give very slightly different energies to corresponding levels. For all but the smallest microcrystals the levels within a band are so closely spaced that they almost form a continuum and so this difference has no consequence.

In the linear model the periodic boundary condition is equivalent to bending the chain of atoms round to form a closed loop. Choosing the coordinates such that the ends of the chain are $x=0$ and $x=NL$, we require $\psi(0)=\psi(NL)$. From (10.3) we see that this requires that $e^{ikNL}=1$, so that the possible values of k are discrete:

$$k=2n\pi/NL, \qquad n=0, \pm 1, \pm 2, \dots \tag{10.13}$$

Furthermore, we find from (10.3), (10.5), (10.6), (10.10) and (10.11) that both ψ and the corresponding energy E are unchanged if k is either increased or decreased by an integral multiple of $2\pi/L$. Thus we may, without losing any of the states, confine k to any chosen interval of length $2\pi/L$, for example

$$-\pi/L<k\leqslant\pi/L. \tag{10.14}$$

As k takes each of the allowed values (10.13) that fall within this interval, we pass through every one of the allowed levels in each band. Thus each band has just N allowed levels, where N is the total number of atoms in the crystal.

This result is not unexpected. When the atoms are very far apart each band consists of just a single level, the energy level of the solitary atom. But the single level corresponds to N different states, because the electron can be attached to any one of the N different atoms. As the atoms are brought together, the N-fold degenerate levels spread out into bands. This is a generalisation of the two-atom problem that we discussed in chapter 6.

The property that the number of levels in each band is equal to the number of atoms in the crystal is special to the simplest crystal structure. In many cases the crystal lattice consists not of a regularly spaced array of single atoms, but rather of an array of *unit cells* that each consist of more than one atom. The model of figure 10.3 represents a crystal in which the unit cell consists of a single atom; in figure 10.5 we show a one-dimensional potential that can be used as a model for a crystal in which the unit cell consists of two atoms. Generally, the number of levels in each energy band is equal to the number of unit cells in the crystal.

Band overlap

Equations such as (10.10) and (10.11), which relate k and E, are called *dispersion relations*. We consider now the solution to these equations, which expresses E as a function of k. It is not possible to obtain an explicit expression for this solution; it is necessary to use either graphical or numerical methods. We discuss the general features of the solution.

A given value of k corresponds to many values of E. The number of values for which $E<0$, corresponding to (10.10), depends on the depth and width of the atomic potential wells, but, for $E>0$, (10.11) always gives an infinite number of E values for a given k. Because (10.10) and (10.11) involve k only through $\cos kL$ the solutions for E are symmetric about the origin in k. According to figure 10.4, the edges of the allowed energy bands occur when $\cos kL = \pm 1$, and if the range of definition of k is chosen to be (10.14) this corresponds to either $k=0$ or $k=\pm\pi/L$. From these considerations, we deduce that the dispersion curves have the form drawn in figure 10.6.

Figure 10.5. One-dimensional model for a crystal having two atoms per unit cell.

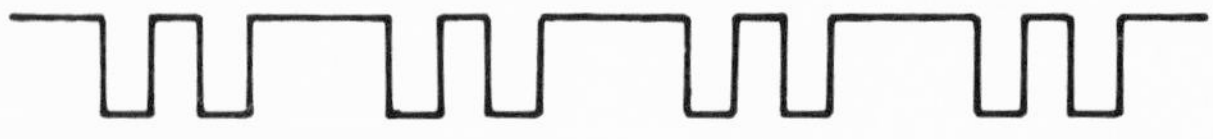

In the one-dimensional model, gaps always occur between the bands, regardless of how high up in energy the bands are. But in the real world of three dimensions the bands may overlap. In three dimensions the Bloch wave function (10.3) becomes

$$\psi(\boldsymbol{r}) = \mathrm{e}^{\mathrm{i}\boldsymbol{k}\cdot\boldsymbol{r}} u_{\boldsymbol{k}}(\boldsymbol{r}), \tag{10.15}$$

so that $\boldsymbol{k}$ is a vector. The dispersion curve becomes a three-dimensional surface in four-dimensional $(E, \boldsymbol{k})$ space. The properties of the crystal may be different in different directions, so that if we draw sections of this surface corresponding to different directions for $\boldsymbol{k}$ the resulting curves need not be the same. Each such section will have an appearance similar to figure 10.6, with gaps between the bands, but the positions of these gaps and their widths may vary with the direction chosen for $\boldsymbol{k}$. As this direction is varied, it may be that the lowest position of the top of a given gap between two bands is lower in energy than the highest position of the bottom of that gap (figure 10.7). If this happens, the bands overlap and there is no gap between them: it is possible to get from one to the other by continuously varying $\boldsymbol{k}$. This occurs in divalent metals and semi-metals, and has important consequences.

Figure 10.6. Dispersion curves corresponding to figure 10.4.

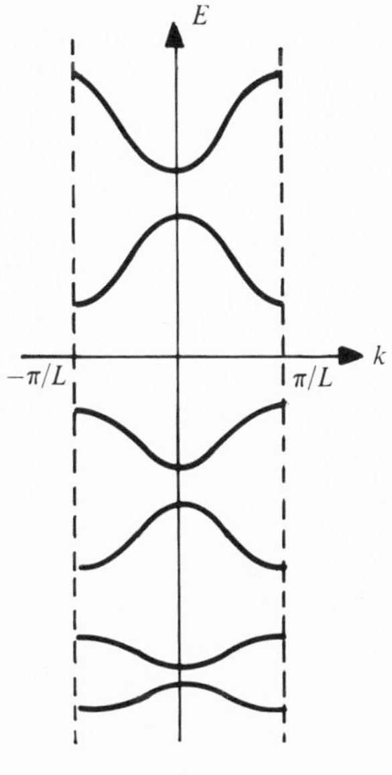

Simple consequences of band structure

The band structure of a material has a profound effect on its physical properties, in particular on its electrical conductivity and its optical properties.

The electrical conductivity takes a wider range of values for different materials than does any other physical property: at room temperature, the ratio of the conductivity of the best conductor to that of the best insulator is about 10^{28}. It is conventional to classify a solid as an insulator, a semiconductor, a semi-metal or a metal, partly according to how good a conductor it is.

We explained in chapter 5 that the Pauli exclusion principle allows at most one electron to be in each quantum-mechanical state. Since the electron has two independent spin configurations (the z-component of its spin has eigenvalues $\pm\frac{1}{2}$), the N levels of a band can accommodate at most $2N$ electrons. At the absolute zero of temperature the electrons will be found in states of as low an energy as possible; that is, the levels are filled up, with two electrons in each, starting from the lowest level of the lowest band.

Suppose that the crystal consists of one type of atom only, each having charge Ze on its nucleus. Although most elements crystallise into forms containing one atom per unit cell, a few of interest to us here form crystals with two atoms per unit cell. In order to keep the discussion general, let the number of atoms per unit cell be m. Then, if the crystal contains N unit cells, it has ZmN electrons when it is electrically neutral. We have said that the maximum number of electrons in any one band is $2N$. Hence at the absolute zero of temperature, $T = 0$, there are three basically different ways in which the bands can be populated, distinguished by whether the number Zm

Figure 10.7. Band overlap.

of electrons per unit cell is even or odd, and if it is even by whether or not there is band overlap. We consider these cases in turn.

Insulators and semiconductors

In these crystals Zm is even and no band overlap occurs. At $T=0$ a number of bands are completely filled with electrons and the remainder are empty. The topmost filled band is called the *valence* band, and the next band above it, which is empty, is called the *conduction* band. If the crystal is placed in a strong electric field, some of the electrons in the valence band may be excited into the conduction band. However, with an electric field of ordinary magnitude the electrons do not acquire enough energy to jump the energy gap between the bands. As all the valence-band states are already filled, the quantum state of the crystal cannot change when such a field is applied. Therefore the field does not cause a net electric current to flow.

At temperature $T>0$ some of the valence-band electrons are thermally excited into the conduction band. In the next chapter we show that the number of electrons that appear in the conduction band varies rapidly with the width Δ of the energy gap and with the temperature T. In an insulator, Δ is so large that the electron concentration in the conduction band, and therefore also the electric current, are negligibly small even up to the melting temperature of the solid. In a semiconductor, Δ is small enough for there to be a significant electron concentration in the conduction band; this increases as T increases.

In the next chapter we explain how this enables the semiconductor to carry an electric current. The electrons that are thermally excited into the conduction band leave unfilled states or *holes* in the levels they previously occupied in the valence band (see figure 10.8). If an electric

Figure 10.8. Excitation of electrons from the top of the valence band of a semiconductor into the bottom of the conduction band.

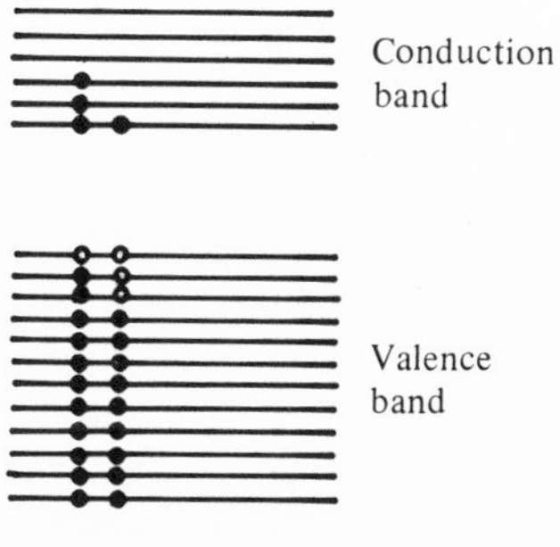

field is applied, the electrons in the conduction band can fairly easily change their states by going to nearby unfilled levels in the conduction band; the quantum state of the crystal changes in such a way that there is a net drift of electrons. Another contribution to this net drift comes from some of the valence-band electrons, which can change their states by occupying the holes in the valence band. Hence the semiconductor carries a current. Such a semiconductor is known as an *intrinsic* semiconductor. In chapter 12 we shall explain that the conductivity of an insulator or an intrinsic semiconductor may be increased by the presence in the crystal of a small concentration of certain impurity atoms; the material is then said to be a *doped* semiconductor.

Semi-metals and divalent metals

Arsenic, antimony and bismuth each have an odd number of electrons per atom, but crystallise into a structure having two atoms per unit cell. Since the total number of electrons per unit cell is even, they would be insulators were it not for the fact that their conduction and valence bands overlap to a small extent (we recall figure 10.7). This means that even at $T = 0$ there are some electrons in the lowest levels of the conduction band, leaving unfilled those states in the valence band that have higher energy. These crystals are therefore fairly good conductors; they are known as *semi-metals*. A similar situation occurs in the divalent metals, of which calcium and strontium are examples. These have an even number of electrons per atom and crystallise into a structure with one atom per unit cell, so that again each has an even number of electrons per unit cell.

Metals

The atoms of the noble metals (copper, silver etc.) and of the alkali metals (sodium, potassium etc.) have an odd number of electrons and one atom per unit cell. Hence their valence band is completely filled and their conduction band half full. The high concentration of conduction-band electrons ensures that metals are good electrical conductors.

For insulators and semiconductors, the magnitude Δ of the energy gap between the valence and conduction bands has an important influence also on the colour of the pure crystal. Consider, for example, diamond, which is one of the crystal forms of carbon. This has $\Delta = 5.6$ eV. A photon in the visible frequency range has energy in the range 1.7 to 3.5 eV, less than Δ, so that an electron in the valence band cannot absorb such a photon. There are not enough conduction-band electrons to give appreciable absorption, so all visible light passes through the crystal. Diamond is therefore colourless. Sulphur, on the other hand, has $\Delta = 2.4$ eV, so that its valence-band electrons can absorb photons of frequency ω such that $\hbar\omega > 2.4$ eV and be knocked into the conduction band. Thus the blue end of the spectrum is absorbed, and the crystal is yellow in appearance. These remarks apply to pure crystals; as we indicate in chapter 12, the introduction of impurities can have an important effect.

Problems

10.1 In the *free-electron model* of a metal the potential is taken to be constant, independent of position in the crystal. What form do the Bloch waves take in the one-dimensional version of the model? Sketch the dispersion curves, analogous to those of figure 10.6. (A more realistic model is the *nearly-free-electron model*, which starts from the free electron model and then uses perturbation theory to introduce a small additional potential around each atomic site.)

10.2 A crystal cube has N atoms arrayed in a simple cubic lattice and its volume V is large. How does (10.14) generalise to this three-dimensional case? How many allowed $\boldsymbol{k}$-vectors are there per unit volume of $\boldsymbol{k}$-space?

If there are Z electrons per atom, what is the largest value of $|\boldsymbol{k}|$ for the states occupied at zero temperature?

10.3 (i) As a model for a one-dimensional crystal, take the potential as

$$V(x) = \sum_{n=-\infty}^{\infty} \left[-\frac{\hbar^2 U}{2m} \delta(x - nL) \right].$$

Investigate the level structure, using a Bloch wave function.

(ii) Explain how to repeat the calculation for the potential

$$V(x) = \sum_{n=-\infty}^{\infty} \left[-\frac{\hbar^2 U_1}{2m} \delta(x - 2nL) - \frac{\hbar^2 U_2}{2m} \delta(x - (2n+1)L) \right],$$

which is a model for a crystal having two types of atom.

10.4 The *tight-binding model* for a crystal is as follows. When the electron is bound to a solitary atom, it has energy E_0 and wave function ϕ, and the binding is tight. In the crystal, the atoms are not very close together. The wave function

$$\psi(x) = \sum_n e^{iknl} \phi(x - nL)$$

clearly satisfies the Bloch condition (E.10) and one can show that it approximately describes a stationary state. By considering the expectation value of the Hamiltonian H, show that

$$E \approx E_0 + 2 \cos kl \int \mathrm{d}x\, \phi^*(x)\, V_0(x)\phi(x-L),$$

where $V_0(x)$ is the potential that would correspond to a solitary atom at the origin. (Notice that $\psi(x)$, as given, is not normalised.)

Deduce that each atomic level E_0 gives a single band of energy levels in the crystal.

10.5 Show that the number of states in a band is unaffected by whether periodic boundary conditions are applied, or instead the electrons are strictly confined within the crystal.

11

Electron motion in crystals

Electron velocity

In the previous chapter we showed that in a perfectly regular crystal the stationary-state electron wave functions are the Bloch waves

$$\Psi_{\boldsymbol{k}}(\boldsymbol{r}, t) = u_{\boldsymbol{k}}(\boldsymbol{r})\, \mathrm{e}^{\mathrm{i}(\boldsymbol{k}\cdot\boldsymbol{r} - E_{\boldsymbol{k}}t/\hbar)}. \tag{11.1}$$

This expression is the same as (10.15), except that we have added the factor $\mathrm{e}^{-\mathrm{i}E_{\boldsymbol{k}}t/\hbar}$ so as to write the time-dependent wave function. The values allowed for the components of the vector $\boldsymbol{k}$ are discrete, but if the crystal is large enough these discrete values are so close together that for many purposes $\boldsymbol{k}$ may be treated as a continuous variable. As we know from the discussion in chapter 4, apart from questions of degeneracy the wave functions corresponding to different values of $\boldsymbol{k}$ are orthogonal, and we may choose the normalisation

$$\int \mathrm{d}^3\boldsymbol{r}\; \Psi^*_{\boldsymbol{k}'}(\boldsymbol{r}, t)\Psi_{\boldsymbol{k}}(\boldsymbol{r}, t) = (\Psi_{\boldsymbol{k}'}, \Psi_{\boldsymbol{k}}) = \delta_{\boldsymbol{k}\boldsymbol{k}'}. \tag{11.2a}$$

Here, the integration is over the volume of the crystal and $\delta_{\boldsymbol{k}\boldsymbol{k}'}$ is zero unless the vectors $\boldsymbol{k}$ and $\boldsymbol{k}'$ are the same, in which case it is equal to 1. From (11.1) and (11.2a),

$$(u_{\boldsymbol{k}'}, u_{\boldsymbol{k}}) = \delta_{\boldsymbol{k}\boldsymbol{k}'}. \tag{11.2b}$$

The Bloch wave function (11.1) satisfies the Schrödinger equation

$$H\Psi_{\boldsymbol{k}} = \mathrm{i}\hbar\dot{\Psi}_{\boldsymbol{k}} \tag{11.3}$$

with

$$H = \hat{p}^2/2m + V(\boldsymbol{r}),$$

where V is the crystal potential. In terms of $u_{\boldsymbol{k}}$, these equations read

$$\hat{h}(\boldsymbol{k})u_{\boldsymbol{k}} = E_{\boldsymbol{k}}u_{\boldsymbol{k}} \tag{11.4}$$

where

$$\hat{h}(\boldsymbol{k}) = (\hat{p} + \hbar \boldsymbol{k})^2/2m + V(\boldsymbol{r}).$$

The Bloch wave function (11.1) is not an eigenfunction of the electron momentum operator $\hat{\boldsymbol{p}} = -\mathrm{i}\hbar\boldsymbol{\nabla}$. But we may calculate the expectation value of the momentum:

$$\langle \boldsymbol{p} \rangle = (\Psi_{\boldsymbol{k}}, -\mathrm{i}\hbar\boldsymbol{\nabla}\Psi_{\boldsymbol{k}}) = (u_{\boldsymbol{k}}, (\hat{\boldsymbol{p}} + \hbar\boldsymbol{k})u_{\boldsymbol{k}}). \tag{11.5}$$

In order to evaluate this, we first write (11.4) with $\boldsymbol{k}$ replaced by some other allowed vector $\boldsymbol{k}' = \boldsymbol{k} + \delta\boldsymbol{k}$. If we approach the continuum limit, $\delta\boldsymbol{k}$ may be chosen to be essentially in any direction and also arbitrarily small in magnitude. If we neglect quadratic and higher powers of $\delta\boldsymbol{k}$, (11.4) gives

$$\hat{h}(\boldsymbol{k})u_{\boldsymbol{k}'} + \delta\boldsymbol{k} \cdot (\partial\hat{h}(\boldsymbol{k})/\partial\boldsymbol{k})u_{\boldsymbol{k}} = E_{\boldsymbol{k}}u_{\boldsymbol{k}'} + \delta\boldsymbol{k} \cdot (\partial E_{\boldsymbol{k}}/\partial\boldsymbol{k})u_{\boldsymbol{k}}. \tag{11.6}$$

We now pre-multiply this equation by $u_{\boldsymbol{k}}^*$ and apply the integration $\int \mathrm{d}^3\boldsymbol{r}$. The first term on the left-hand side becomes

$$(u_{\boldsymbol{k}}, \hat{h}(\boldsymbol{k})u_{\boldsymbol{k}'}) = E_{\boldsymbol{k}}(u_{\boldsymbol{k}}, u_{\boldsymbol{k}'}),$$

where we have again used (11.4). This vanishes because of (11.2*b*). The first term on the right-hand side of (11.6) similarly gives zero contribution. The other terms in (11.6) give

$$\delta\boldsymbol{k} \cdot \left(u_{\boldsymbol{k}}, \frac{\partial\hat{h}(\boldsymbol{k})}{\partial\boldsymbol{k}} u_{\boldsymbol{k}} \right) = \delta\boldsymbol{k} \cdot \frac{\partial E_{\boldsymbol{k}}}{\partial\boldsymbol{k}}, \tag{11.7}$$

where again (11.2*b*) has been used to simplify the right-hand side. Because the direction of $\delta\boldsymbol{k}$ is arbitrary, (11.7) remains true if we cancel the factor $\delta\boldsymbol{k}$ on each side. From the definition of the operator $\hat{h}(\boldsymbol{k})$ in (11.4),

$$(\partial\hat{h}(\boldsymbol{k})/\partial\boldsymbol{k}) = (\hbar/m)(\hat{\boldsymbol{p}} + \hbar\boldsymbol{k}). \tag{11.8}$$

Thus (11.5), (11.7) and (11.8) give

$$\langle \boldsymbol{p} \rangle = (m/\hbar)(\partial E_{\boldsymbol{k}}/\partial\boldsymbol{k}) \tag{11.9a}$$

or

$$\boldsymbol{v} = (1/\hbar)(\partial E_{\boldsymbol{k}}/\partial\boldsymbol{k}), \tag{11.9b}$$

where $\boldsymbol{v} = \langle \boldsymbol{p} \rangle / m$ is the expectation value of the velocity of the electron.

In classical mechanics, the direction and magnitude of the electron velocity would continually be changing, because of collisions between the electron and the atoms of the crystal. The result (11.9*b*) shows that in quantum mechanics $\boldsymbol{v}$ is constant in both direction and magnitude. This result is valid only for a perfectly regular crystal lattice.

In the one-dimensional model described in the previous chapter, the graphs of E_k plotted against k are as drawn in figure 10.6. We reproduce these in figure 11.1, together with the graphs of $\partial E_k/\partial k$ and $\partial^2 E_k/\partial k^2$, for the three lowest bands. Notice that the maximum value of $v = \hbar^{-1}\,\partial E_k/\partial k$ increases as we go up to higher bands. This is because the width of the bands increases with increasing band number, while the range of definition of k (given in (10.14)) remains fixed.

Motion in an external electric field

The Bloch wave functions (11.1) describe stationary states, corresponding to no variation with time of the probability density at each point in the crystal. If we wish to consider the motion of an electron in the presence of an external electric field, it is more appropriate to use a localised wave packet. This is a superposition of stationary-state wave functions:

$$\Psi(\boldsymbol{r}, t) = \int d^3\boldsymbol{k}\, N(\boldsymbol{k})\Psi_{\boldsymbol{k}}(\boldsymbol{r}, t), \qquad (11.10)$$

where the function $N(\boldsymbol{k})$ determines the shape and spread of the wave

Figure 11.1. Plots of E_k, $\partial E_k/\partial k$ and $\partial^2 E_k/\partial k^2$ against k for the three lowest bands of figure 10.6.

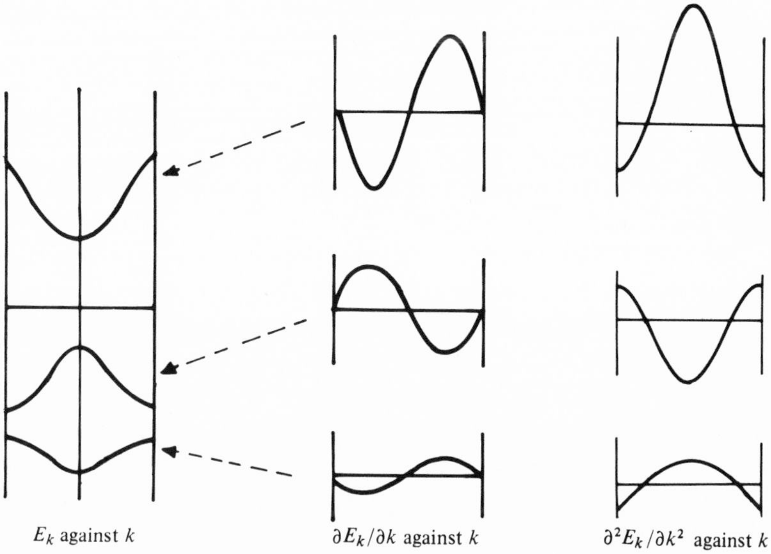

packet, and $\int \mathrm{d}^3\boldsymbol{k}$ means integration over each of the components of the vector $\boldsymbol{k}$. Because each Bloch wave (11.1) satisfies the time-dependent Schrödinger equation (11.3), so also will any superposition (11.10). However, (11.10) is not a stationary state.

We shall suppose that $N(\boldsymbol{k})$ is sharply peaked at some value of $\boldsymbol{k}$ and that it falls rapidly to zero outside the immediate vicinity of this value. This means that each component of the wave packet corresponds to approximately the same value $\boldsymbol{v} = \partial E_{\boldsymbol{k}}/\partial \boldsymbol{k}$ of the expectation value of the electron velocity. We explained in chapter 4 that a narrow spread in $\boldsymbol{k}$ corresponds to a wide spread in $\boldsymbol{r}$. If we had, instead, chosen a wave packet whose spread in $\boldsymbol{r}$ was less than the dimension of a unit cell of the crystal, the expectation value of the electron velocity would have varied as the wave packet moved across each unit cell. This variation may be traced back to the variation over the unit cell of the factor $u_{\boldsymbol{k}}(\boldsymbol{r})$ in the Bloch wave function (11.1). But for a wave packet that spans many unit cells, the variation of $u_{\boldsymbol{k}}(\boldsymbol{r})$ is averaged out and the time variation of $\boldsymbol{v}$ is negligible.

That is the situation in the absence of an external electric field. In problem 4.3 we proved Ehrenfest's theorem (4.14), that a wave packet moves like a classical particle. Hence in an electric field $\boldsymbol{\mathscr{E}}$ the vector $\boldsymbol{k}$ about which $N(\boldsymbol{k})$ is peaked changes with time in such a way that the rate of change of the electron energy is equal to the rate of work done by the field:

$$\mathrm{d}E_k/\mathrm{d}t = -e\boldsymbol{\mathscr{E}} \cdot \boldsymbol{v}. \qquad (11.11a)$$

But $\mathrm{d}E_{\boldsymbol{k}}/\mathrm{d}t = \dot{\boldsymbol{k}} \cdot \partial E_{\boldsymbol{k}}/\partial \boldsymbol{k} = \hbar\dot{\boldsymbol{k}} \cdot \boldsymbol{v}$, and so

$$\dot{\boldsymbol{k}} = -(e/\hbar)\boldsymbol{\mathscr{E}}. \qquad (11.11b)$$

This result holds provided that the necessary empty electron states exist so as to allow $\boldsymbol{k}$ to change.

When $\boldsymbol{\mathscr{E}}$ is constant, $\boldsymbol{k}$ increases uniformly with time in the direction opposite to $\boldsymbol{\mathscr{E}}$. At whatever point of the dispersion curve $\boldsymbol{k}$ starts, it moves in this direction until it reaches the limit of the range of definition of the vector $\boldsymbol{k}$. In the one-dimensional model, this is the point G in figure 11.2, corresponding to $k = \pi/L$. Because the states with $k = \pi/L$ and $k = -\pi/L$ coincide, the representative point now reappears at the point G' in figure 11.2 and then it again moves across towards G. From figure 11.1 we see that this means that the electron velocity $v = \hbar^{-1}\,\partial E_k/\partial k$ oscillates back and forth as the cycle is repeated. This means that in real space the electron oscillates back and forth also.

We have assumed that no band overlap occurs, as is true in the one-dimensional model. When there is band overlap, even a weak field may induce the electron to make a transition from one band to another, and then the motion is more complicated.

Electric current

In a full band, all the allowed representative points along the corresponding dispersion curve are occupied and the exclusion principle therefore forbids the distribution of these points to change when a field is applied. For each electron with velocity $\boldsymbol{v}$ there is another with velocity $-\boldsymbol{v}$, and so the total current due to all the electrons in a full band vanishes.

But suppose that the conduction band is half full and is separated from the band above it by a gap. Let the dispersion curve for the conduction band have the shape drawn in figure 11.2; if, instead, it is the other way up like the middle band in figure 11.1, the discussion must be modified accordingly. The representative points for the electrons, which are initially in the lower half of the band, will be strung out along the (v, k) curve like a chain of beads. When there is no electric field the points are distributed symmetrically about $k = 0$, as in figure 11.3(a), so that the total current is zero. When a field is applied towards the left, the representative points begin to move to the right, as shown in figure 11.3(b). The vector sum of the velocities of the collection of electrons is now non-zero, and so a net electric current flows in the direction of the electric field (remember that the electrons have negative charge). When the electrons reach the right-most edge

Figure 11.2. Motion of the representative point in the presence of an electric field $\mathscr{E}$. When the point reaches G, it reappears at G'.

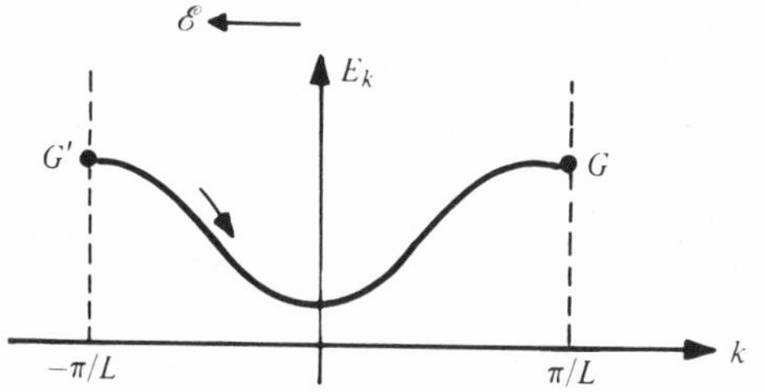

of the range of definition of k they reappear at the left, figure 11.3(c), until in figure 11.3(d) the sum of the velocities has changed sign. An electric current now flows in the reverse direction. Subsequently, it changes direction again, and so on.

This rather surprising result, that the direction of the electric current oscillates back and forth with time, does not apply if the conduction band overlaps the next higher band, and if all the higher bands similarly overlap each other. Then the electric current increases steadily, remaining in the direction of the applied field.

In either case, the steady state described by Ohm's law is not found; the crystal has no electrical resistance. This is because the discussion has assumed that the crystal lattice is perfectly regular. In practice this is never exactly true. Even at the absolute zero of temperature the nuclei are not completely anchored at their lattice sites, but oscillate with zero-point energy (see the discussion of the harmonic oscillator in chapter 3). In addition, the exact periodicity of the lattice is disturbed, in nearly all crystals, by the presence of imperfections such as geometrical *dislocations* and concentrations of *impurity* atoms.

Imperfections and impurities apart, the main source of resistance in a solid is the thermal vibration of the atoms about their lattice sites.

Figure 11.3. The velocity of conduction band electrons (a) when there is no electric field; (b), (c) and (d) with an electric field $\mathscr{E}$, at successive times.

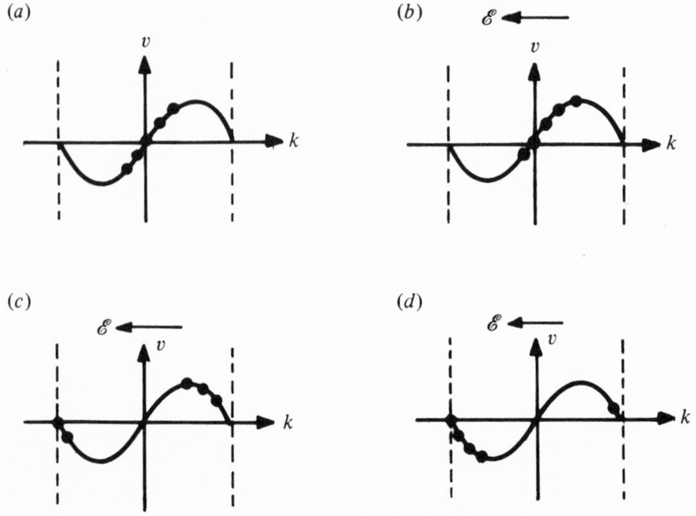

Because the displaced atoms exert forces on one another, their vibrations are correlated. This correlation allows an analysis in terms of *lattice waves*; these travel through the crystal with speeds of the order of the speed of sound. Just as the energy of an electromagnetic wave is quantised, so is the energy of a lattice wave; the quantum of energy is called a *phonon*. Phonons behave in many respects like particles, as do photons.

As the electrons move through the crystal, they repeatedly collide with phonons. As the temperature T increases, the collisions become more frequent and more violent. When an electric field is applied to the crystal, the representative points illustrated in figure 11.3(a) begin to move according to (11.11), but this motion is opposed by the phonon scattering. The scattering with phonons tends to prevent the electrons from increasing their energy and so from rising into the upmost parts of the conduction band. A steady state is rapidly reached: the distribution of representative points becomes static, displaced asymmetrically from its original position as in figure 11.3(b). A steady electric current then flows.

Effective mass and holes

We have seen that when a force is applied to an electron in a perfectly regular crystal, the electron may sometimes move in the direction opposite to the force. We now show that, as a result of the presence of the crystal lattice, the electron moves as if it had a mass m^* different from its real mass m. This effective mass m^* varies with the energy $E_{\mathbf{k}}$, and in a part of each band it is negative.

For simplicity, consider the one-dimensional case. From (11.9b),

$$\frac{\mathrm{d}v}{\mathrm{d}t}=\frac{\mathrm{d}}{\mathrm{d}t}\left(\hbar^{-1}\frac{\partial E_k}{\partial k}\right)=\hbar^{-1}\frac{\partial^2 E_k}{\partial k^2}\dot{k}. \tag{11.12}$$

If we now use (11.11b), this gives

$$m^*\,\mathrm{d}v/\mathrm{d}t=-e\mathscr{E} \tag{11.13}$$

with

$$\frac{1}{m^*}=\frac{1}{\hbar^2}\frac{\partial^2 E_k}{\partial k^2}. \tag{11.14}$$

The relation (11.13) is just Newton's law for the motion of the electron, but with an effective mass m^*.

We have plotted $\partial^2 E_k/\partial k^2$ for the one-dimensional model in figure 11.1; the effective mass m^* is proportional to $(\partial^2 E_k/\partial k^2)^{-1}$. For each of the bands, m^* is negative near the top of the band.

This means that the acceleration of an electron near the top of a band is opposite to the applied force. Suppose that a particular band is nearly full, with a few levels near the top of the band unfilled. This occurs in a semiconductor, where at zero temperature the valence band is filled and the conduction band is empty, but at non-zero temperature a few of the electrons from the top of the valence band are thermally excited into the conduction band. It is convenient to regard the unfilled states as fictitious particles, called *holes*. If the levels were filled with electrons, these would have negative charge and negative effective mass. The holes correspond to an absence of these electrons and so they have positive charge and positive mass. The dynamics of the charge carriers in the nearly filled band may then be described *either* in terms of the large number of electrons in the band *or* in terms of the small number of holes.

We illustrate this by considering the case in which a single electron is missing from an otherwise full band. When an electric field is applied, the total electric current contributed by the band is

$$J = -e \sum_{i \neq j} v_i, \tag{11.15a}$$

where the summation extends over all the states i except for the unfilled state j. But when the band is full there is no net electric current, because for each state corresponding to velocity v there is another corresponding to velocity $-v$. Hence

$$0 = -e \sum_{i} v_i,$$

where the summation extends over the whole band, and so (11.15a) may also be written as

$$J = e v_j. \tag{11.15b}$$

This is just the current that would be contributed by a solitary positively charged carrier with velocity v_j equal to that which an electron occupying the level j would have had. Similarly, the rate of change of momentum of all the electrons in the band is

$$\frac{\mathrm{d}P}{\mathrm{d}t} = \sum_{i \neq j} m_i^* \frac{\mathrm{d}v_i}{\mathrm{d}t}. \tag{11.16a}$$

But when the band is filled, the exclusion principle forbids its overall quantum state from changing, so that then the rate of change of its total momentum vanishes:

$$0=\sum_i m_i^* \frac{dv_i}{dt}.$$

Hence (11.16*a*) may also be written as

$$dP/dt=(-m_j^*)\, dv_j/dt. \tag{11.16b}$$

So the rate of change of momentum also may be calculated by ignoring the electrons, and assigning to the unfilled states j fictitious particles of effective mass $(-m_j^*)$ and velocity v_j.

Thermal excitation

In a metal or semi-metal at zero temperature, all states in the conduction band with energies up to some maximum value ζ_0 are filled, and the rest are empty. At temperatures $T>0$ the distribution of occupied states does not end abruptly at ζ_0. As a result of thermal excitation, electrons are found with energies greater than ζ_0; the distribution tails off smoothly beyond energy ζ_0. In a semiconductor or insulator at $T=0$, there are no electrons in the conduction band and no holes in the valence band. At $T>0$ some electrons are excited into the conduction band and leave behind holes in the valence band.

To study the effects of temperature it is rarely necessary to consider the details of the collision processes that cause electron excitation. In a solid in thermal equilibrium, the electrons are distributed in the available energy levels in accordance with the predictions of statistical mechanics. Although the electrons continually change levels as a result of collisions, statistical mechanics gives their expected average distribution among the levels independently of the details of the collisions. Because the effects of the exclusion principle are important, the classical Maxwell–Boltzmann distribution (9.10) is not applicable; instead, the appropriate distribution is the *Fermi–Dirac distribution*. If the level E_i has degeneracy g_i (with a minimum value $g_i=2$ when spin is taken into account), the average number of electrons expected in that level is

$$n_i=\frac{g_i}{e^{(E_i-\zeta)/k_B T}+1}. \tag{11.17}$$

Here, the parameter ζ is called the *Fermi energy* or, more correctly, the *chemical potential.* It is independent of energy, but it varies with the temperature T. It is determined in terms of the total number N of electrons by the implicit equation

$$N = \sum_i n_i = \sum_i \frac{g_i}{e^{(E_i - \zeta)/k_B T} + 1}. \tag{11.18}$$

This usually has the effect that ζ varies as log $k_B T$, so that it varies rather slowly with T. In practice, the variation between $T=0$ and room temperature is almost negligible.

The Fermi–Dirac distribution (11.17) is derived in any textbook on statistical mechanics. It has the property that at zero temperature

$$\begin{aligned} n_i &= g_i \qquad E_i < \zeta_0 \\ n_i &= 0 \qquad E_i > \zeta_0 \end{aligned} \tag{11.19}$$

where ζ_0 is the value of ζ at $T=0$. For temperatures $T>0$ the distribution has the shape drawn in figure 11.4; in this figure we have supposed that all the degeneracies g_i have the same value g. For large values of E_i, such that $(E_i - \zeta) \gg k_B T$, the exponential in the denominator of (11.17) dominates and

$$n_i \approx C g_i \, e^{-E_i/k_B T}, \tag{11.20}$$

where $C = e^{\zeta/k_B T}$. So for energies far above the Fermi energy we again have the classical Maxwell–Boltzmann distribution (9.10). This is because the average number n_i of electrons in each level is then very small, so that the exclusion principle has little effect.

Figure 11.4. The Fermi–Dirac distribution at temperatures $T>0$, for the case in which all the levels have the same degeneracy g.

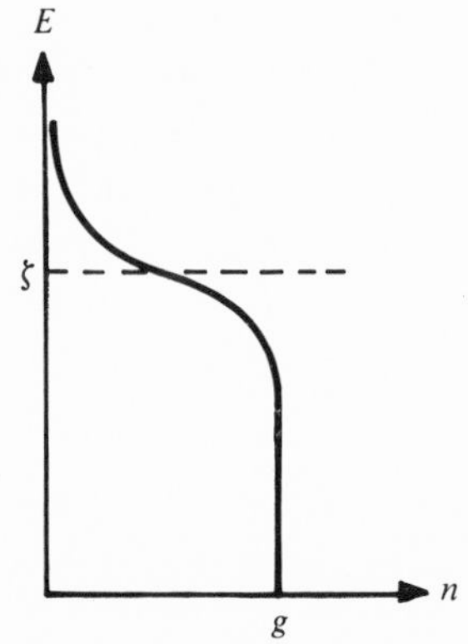

In any bound quantum system of finite size the allowed values of E_i are discrete, but for a large system they are usually so close together that it is an excellent approximation to treat E_i as a continuous variable. Let $\rho(E)\,\mathrm{d}E$ be the number of states in the energy range E to $E+\mathrm{d}E$. The expected average number of electrons in this range is then, as the continuum limit of (11.17),

$$n(E)\,\mathrm{d}E=\frac{\rho(E)\,\mathrm{d}E}{\mathrm{e}^{(E-\zeta)/k_{\mathrm{B}}T}+1}. \tag{11.21}$$

The equation that determines ζ is the analogue of (11.18):

$$N=\int\frac{\rho(E)\,\mathrm{d}E}{\mathrm{e}^{(E-\zeta)/k_{\mathrm{B}}T}+1}. \tag{11.22}$$

The average number density $n(E)$ given by (11.21) is the product of $\rho(E)$ and a factor that has the shape drawn in figure 11.4 (with g set equal to 1). In the energy bands, $\rho(E)$ is a function that depends on the nature of the crystal, whereas in the gaps $\rho(E)=0$. Unless the temperature T is very high, the Fermi level ζ for a metal or semi-metal lies in the conduction band, while for an insulator or a semiconductor it lies somewhere in the energy gap between the valence band and the conduction band. We demonstrate this last result. In an insulator or an intrinsic semiconductor, there are just enough electrons to fill all the levels up to the top of the valence band:

$$N=\int_{-\infty}^{E_{\mathrm{v}}}\mathrm{d}E\,\rho(E), \tag{11.23}$$

where E_{v} is the energy at the top of the valence band. But from (11.22) evaluated at $T=0$,

$$N=\int_{-\infty}^{\zeta_0}\mathrm{d}E\,\rho(E). \tag{11.24}$$

That is, at $T=0$ all the levels up to the Fermi energy ζ_0 are filled, and all those above it are empty. For (11.23) and (11.24) to be consistent, $\rho(E)$ must vanish for values of E between E_{v} and ζ_0, so that ζ_0 must be in the energy gap. As T is increased ζ changes slowly, but remains in the energy gap until T becomes much greater than room temperature.

Pair excitation in intrinsic semiconductors

There are two obvious ways in which an electron in an intrinsic semiconductor can be knocked from the valence band into the conduction band, so producing an electron–hole pair of current carriers. The first way is to bombard the crystal with photons whose frequency ω is such that $\hbar\omega > \Delta$, where Δ is the magnitude of the energy gap between the valence band and the conduction band. The rate of pair production increases with the intensity of the light, and so when a steady electric field is applied the electric current also increases with the light intensity. This is the principle of the *photoelectric cell.*

Secondly, when $T>0$ pair excitation occurs because of collisions between electrons and phonons. The number of pairs so excited can be calculated from the Fermi–Dirac distribution (11.21). Let the number of electrons excited into the conduction band, or into even higher bands, be N_e. Because the exponential in the denominator of the Fermi–Dirac distribution grows rapidly as E increases, most of these excited electrons will actually be near the bottom of the conduction band, and most of the holes that they leave behind will be near the top of the valence band. Hence, from what we have said earlier in this chapter, both the excited electrons and the holes will nearly all have positive effective mass m^*.

If again E_v is the energy at the top of the valence band, and Δ is the energy gap, the Fermi–Dirac distribution (11.21) gives the total number of electron-hole pairs as

$$N_e = \int_{E_v+\Delta}^{\infty} dE \frac{\rho(E)}{e^{(E-\zeta)/k_B T}+1}. \tag{11.25}$$

On the other hand, the total number of electrons in the valence band and in lower bands is

$$\int_{-\infty}^{E_v} dE \frac{\rho(E)}{e^{(E-\zeta)/k_B T}+1}. \tag{11.26}$$

This must be equal to $(N - N_e)$, where the total number N of electrons is given by (11.23). Thus

$$\int_{E_v+\Delta}^{\infty} dE \frac{\rho(E)}{e^{(E-\zeta)/k_B T}+1} = \int_{-\infty}^{E_v} dE\, \rho(E) \left[1 - \frac{1}{e^{(E-\zeta)/k_B T}+1}\right]$$
$$= \int_{-\infty}^{E_v} dE \frac{\rho(E)}{e^{-(E-\zeta)/k_B T}+1}. \tag{11.27}$$

This is the equation that determines $\zeta(T)$.

For purposes of illustration, consider the special case where the level density $\rho(E)$ has the same shape near the bottom of the conduction band as at the top of the valence band (figure 11.5). To a reasonable approximation, this is true in silicon, germanium and many other semiconductors. Then near the energy gap

$$\rho(E)=\rho(E_v+2\Delta-E). \qquad (11.28)$$

Most of the contribution to the integral on the left-hand side of (11.27) comes from near the bottom end of the integration, and most of the contribution to the integral on the right-hand side comes from the top end, so we do not need to know about $\rho(E)$ elsewhere. One may readily check that (11.27) is satisfied by

$$\zeta=E_v+\tfrac{1}{2}\Delta. \qquad (11.29)$$

So in this special case ζ is independent of T and lies exactly in the middle of the energy gap.

Inserting (11.29) into (11.25) and evaluating it for the case $k_BT \ll \Delta$, so that the exponential in the denominator dominates, we have

$$N_e=\int_{E_v+\Delta}^{\infty} dE\,\rho(E)\,e^{-(E-E_v-\frac{1}{2}\Delta)/k_BT}. \qquad (11.30)$$

Since most of the contribution to this integral arises from values of E near the lower end of the integration, we see that approximately

$$N_e \propto e^{-\Delta/2k_BT}. \qquad (11.31)$$

At room temperature, $k_BT \approx \frac{1}{40}$ eV. For silicon or germanium, the energy gap Δ is about 1 eV, so that the exponential is very small indeed at room temperature. However, the total number N of electrons is very large, so that the exponential is multiplied by a very large number, and at room temperature the conductivity is fairly good. But the

Figure 11.5. Model in which the state-density function $\rho(E)$ has the same shape at the top of the valence band as at the bottom of the conduction band

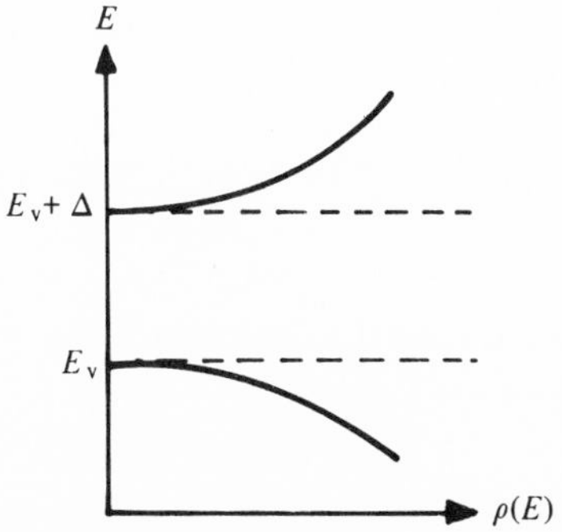

exponential decreases rapidly as T decreases, and at around 0 °C the conductivity of silicon or germanium is only a seventh of its value at room temperature (remember that $T=0$ corresponds to -273 °C). Electronic devices that contain semiconductors frequently incorporate feedback loops in their circuits in order to compensate for changes of temperature.

Problems

11.1 Consider the free-electron model for a crystal (problem 10.1), defining the constant potential to be zero inside the crystal.
(i) In the case of a one-dimensional lattice, show that

$$\rho(E)=(NL/\pi\hbar)(m/2E)^{1/2}.$$

(ii) In the case of a simple square lattice of area A, show that

$$\rho(E)=(Am/\pi\hbar^2).$$

(iii) In the case of a simple cubic lattice of volume V, show that

$$\rho(E)=(V/\pi^2\hbar^3)(2m^3E)^{1/2}.$$

11.2 An atomic monolayer is formed by evaporating a metal onto a flat substrate. As a model for the behaviour of the electrons in the metal film, suppose that the potential $V(\mathbf{r})$ vanishes for $0<z<a$, where a is the atomic diameter, and is infinite otherwise. Hence the electrons travel freely in the film. How does the density of states $\rho(E)$ vary with E?

11.3 In the free-electron model of a metal, the electrons behave as a set of independent particles obeying the Fermi–Dirac distribution. An electron having momentum component normal to the metal surface greater than a critical value $(2mV_0)^{1/2}$ can escape from the surface; this is called *thermionic emission.* In practice $V_0 \gg \zeta + k_B T$. If a collecting electrode is placed near the metal surface and is maintained at a sufficiently positive potential relative to the metal, it will collect all the electrons that are emitted. Show that the current per unit area is

$$\frac{emk_B^2T^2}{2\pi^2\hbar^3}\,e^{(\zeta-V_0)/k_BT}.$$

11.4 As a model of a semiconductor, suppose that there are N electron states all having the same energy $-\Delta$, and that these are all filled at zero temperature. (That is, the valence band has zero width.) The conduction band is a continuum of states of energy $E \geqslant 0$, with density given by

$$\rho(E)=A\sqrt{E}$$

with A a constant. Show that the number of electrons in the conduction band at temperature T is

$$n=N[1+e^{(\Delta+\zeta)/k_BT}]^{-1}$$

where, if T is so small that $e^{\zeta/k_BT} \ll 1$ and that $n \ll N$,

$$\zeta \approx -\tfrac{1}{2}\Delta+\tfrac{1}{2}k_BT\log[2N/A(k_BT)^{3/2}\pi^{1/2}].$$

12

Transistors

Impurities

We have so far described the properties of intrinsic semiconductors, in which the crystal lattice is perfectly regular. But the electrical properties of semiconductors, unlike metals or semi-metals, are drastically affected by the addition of small traces of impurity atoms. Observable effects occur with impurity concentrations as low as a few parts in 10^8, and increasing the impurity concentration to one part in 10^5 can increase the conductivity by as much as a factor of 10^3 at room temperature and 10^{12} at liquid-helium temperatures. The semiconductor is said to be *doped* with impurity atoms.

In order to study the effect of doping, we first consider a crystal consisting of a periodic array of one type of atom, except that just one of the atoms has been replaced by an atom of a different type. As a one-dimensional model of this situation, we take the same infinite chain of square wells as in chapter 10, but with one of the wells having a different depth, U_1 say (figure 12.1).

We recall that for the perfectly regular crystal, the stationary state solutions are Bloch waves of the form (10.3). These have the property that when the required continuity conditions on the wave function are satisfied at the two edges of one of the potential wells, they are automatically satisfied also at the edges of all the other wells. For this reason, we might guess that in the case of the potential of figure 12.1, the stationary-state wave function is again composed of Bloch waves in

Figure 12.1. One-dimensional model for a crystal with a single impurity atom.

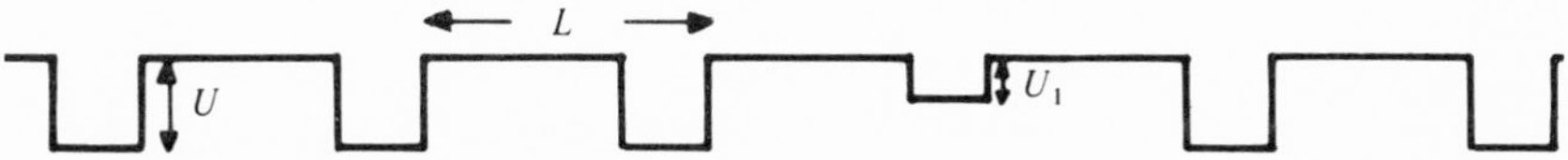

the parts of the crystal where the lattice is perfectly regular, but that it has a different form in the neighbourhood of the impurity.

The Bloch wave $e^{ikx}u_k(x)$, with k real, represents a wave travelling from left to right. If such a wave is incident on the impurity, part of it is reflected, and the remaining part is transmitted across the impurity (compare (3.24) and (3.25)). Hence the total wave function to the left of the impurity atom is

$$e^{ikx}u_k(x)+R\,e^{-ikx}u_{-k}(x), \tag{12.1}$$

and to the right of the impurity atom it is

$$T\,e^{ikx}u_k(x). \tag{12.2}$$

Here, R and T are constants. Inside the potential well that represents the impurity atom, the solution to the time-independent Schrödinger equation is

$$B\,e^{i\gamma x}+C\,e^{-i\gamma x}$$

where (12.3)

$$\hbar^2\gamma^2/2m=E+U_1.$$

The boundary conditions at the edges of the wells corresponding to the *host* atoms, that is, at the edges of the wells other than the impurity, determine E in terms of k. The relationship is exactly as in chapter 10: there is the same band structure as if the impurity were not there. The constants R, T, B and C are determined by the four conditions that ψ and ψ' be continuous at both edges of the impurity well.

There is also a different type of solution, in which the electron is more or less localised in the vicinity of the impurity atom. In this solution, the wave function still satisfies the Bloch condition (E.10), except near to the impurity, but k is pure imaginary. Take the origin $x=0$ to be in the middle of the impurity well. To the left of the impurity the wave function is

$$e^{\kappa x}v_1(x) \qquad (\kappa>0), \tag{12.4}$$

and to the right of the impurity it is

$$e^{-\kappa x}v_2(x), \tag{12.5}$$

where the functions $v_1(x)$ and $v_2(x)$ are both periodic. Because of the exponential factors, the wave function goes to zero as $x\to\pm\infty$, so the electron is *trapped* in the neighbourhood of the impurity.

By making $v_1(x)$ and $v_2(x)$ periodic, we again ensure that when we satisfy the required continuity conditions at the edges of any single well on either side of the impurity, they are automatically satisfied at the edges of the well corresponding to all the host atoms. For a given κ, the functions v_1 and v_2 are determined in a way precisely similar to the determination of $u_k(x)$ in chapter 10. Each contains a multiplicative constant that remains to be fixed and that determines its normalisation. Since the complete potential is symmetric about $x=0$, the solutions fall into two classes. One class has positive parity and the other negative parity, $\psi(x)=\pm\psi(-x)$. For definiteness, consider the positive-parity solutions (the negative-parity ones may be discussed similarly). For these $v_1(x)=+v_2(-x)$, so that the pair of functions v_1 and v_2 now contains one multiplicative constant that remains to be fixed, λ say. Inside the impurity well, the positive-parity solution is

$$B(e^{i\gamma x}+e^{-i\gamma x}), \tag{12.6}$$

where γ is defined in (12.3). Imposing a normalisation condition on the overall wave function, for example $(\psi, \psi)=1$, determines the constant B in terms of λ. There are two more continuity conditions to be satisfied: that ψ and ψ' are continuous at one edge of the impurity well (the corresponding conditions at the other edge are then automatically satisfied, because of the even-parity condition $\psi(x)=\psi(-x)$). Since there are only two unknowns, λ and E (which determines γ and κ), we expect a discrete set of solutions.

Thus, in addition to the usual band structure associated with the host atoms, there are discrete trapped-electron levels resulting from the impurity. This result continues to apply when further impurity atoms are substituted for host atoms, provided that their concentration remains so low that they are far enough apart for there to be little overlap of the trapped-electron wave functions. Each impurity level has degeneracy equal to twice the number of impurity atoms (the factor of two corresponding to the two possible electron spin states). But if their concentration becomes greater than about one part in 10^3, the impurity levels begin to spread out into bands, in the same way as the levels of the host atoms spread into bands when a large number of host atoms are brought together to form a crystal.

n- *and* p-*type semiconductors*

Suppose that a crystal of silicon is doped with a few parts per million of phosphorus. Silicon has a valence of 4, but phosphorus has a valence of

5. So at each lattice site where a phosphorus atom replaces a silicon atom there are five valence electrons, whereas only four are needed to form the covalent bonds that anchor the atom to its nearest-neighbour host atoms (figure 12.2(*a*)). The extra electron either occupies one of the degenerate discrete impurity levels, in which it is trapped near an impurity atom, or it is found in the conduction band. It turns out that the lowest impurity level lies just below the bottom of the conduction band (figure 12.2(*b*)). So at zero temperature the fifth valence electron is in this level, because it is the lowest level available to it. But when $T>0$ the probability is high that the electron is thermally excited into the conduction band. In terms of the Fermi–Dirac distribution, this is because the density of states $\rho(E)$ is much greater in the conduction band, where there are very many levels close together, than in the impurity level.

Figure 12.2. (*a*) A silicon crystal doped with the donor impurity phosphorus, with (*b*) the impurity level just below the conduction band.

(*a*)

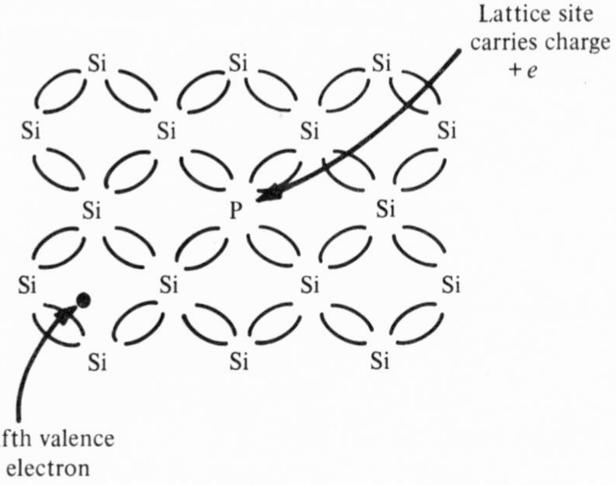

(*b*)

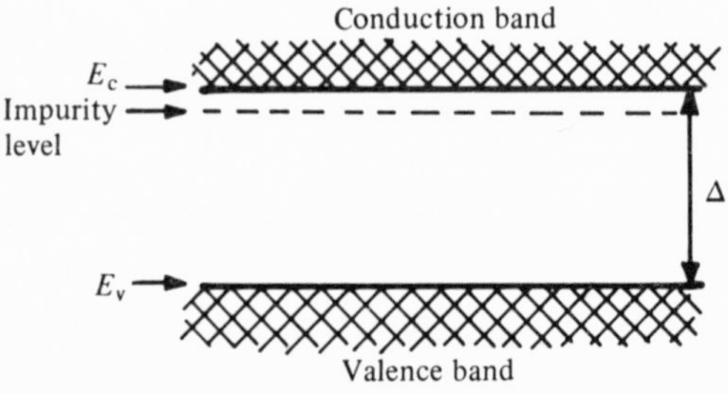

The effect of doping silicon with phosphorus is therefore to provide an excess of electrons in the conduction band, that is an excess of negative carriers which can contribute to an electric current. A material doped in this way is called an *n-type* semiconductor; it is said to be doped with *donor* atoms.

Alternatively, the silicon crystal may be doped with atoms that have a valence of 3, such as boron atoms. The lowest impurity level is then found to be just above the top of the valence band. The impurity atoms each bring with them one electron less than is needed to saturate the nearest-neighbour covalent bonds at the sites that they occupy. At temperatures $T > 0$ electrons are thermally excited from the valence band into the degenerate impurity level, being trapped near the impurity atoms and so providing the missing bonds. They leave behind holes in the valence band. These are mobile and are positive carriers which can contribute to an electric current. The material is known as a *p-type* semiconductor; it is said to be doped with *acceptor* atoms.

Impurities and crystal colour

The presence of impurity atoms can have an important effect on the colour of a crystal. For example, in pure aluminium oxide the energy gap between the valence band and the conduction band is so large that visible light is not absorbed; the crystal is transparent. If a small quantity of chromium is added, say $\frac{1}{2}\%$, there are two relatively narrow bands of impurity levels. These are situated in the energy gap, in such a position that photons corresponding to green and yellow light are absorbed. So now the crystal appears red; in fact, it is ruby. If instead titanium is added, the position of the impurity levels is such that red light is absorbed. The crystal now appears blue; it is sapphire.

Semiconductor junction

Consider two pieces of a given semiconductor, one doped with donor atoms and the other with acceptor atoms. Suppose that each piece has a face that is accurately plane, and imagine bringing the two pieces into contact at their plane faces. This forms a *pn-junction.* (In practice, rather than pressing together two separate pieces of crystal, the junction is manufactured from a single piece of host crystal by varying the doping in different parts of it as the crystal is grown. This produces a transition region between the p-part and n-part that is typically about 1 μm in width. In our discussion, we imagine that the transition is abrupt.)

At the instant when the two pieces are brought together the p-type material has an excess of holes in the valence band, compared with the n-type material. Similarly, the n-type material has an excess of electrons in the conduction band, compared with the p-type material. See figure 12.3(*a*). After the contact is made, it is energetically favourable for some of the excess electrons in the conduction band of the n-type material to cross to the p-type material and annihilate some of the holes there. As they do so, a net negative charge is built up in the p-type material, and a net positive charge in the n-type material. Thus an electrostatic potential is set up, and this eventually stops the flow of further electrons across the junction. The shape of this electrostatic potential ϕ is drawn in figure 12.3(*c*). It causes an additional energy $-e\phi$ to be superimposed on the level structure, so that when dynamical equilibrium is reached the band structure becomes as drawn in figure 12.3(*b*). Typically, the region where the electrostatic potential ϕ varies appreciably, so that there is an appreciable electric field $-\nabla\phi$, is of width 10 μm.

Suppose that heat is applied to the junction, so that extra electron–hole pairs are created there by thermal excitation. The electric field $-\nabla\phi$ drives the electrons towards the n-type material and the holes towards the p-type material. So if the two sides of the crystal are joined to an external circuit, the effect of the heat is to drive a current through the crystal from the n-type side to the p-type side, and round the circuit.

In an exactly similar way, a current is generated if light is shone on the junction, so that the absorbed photons create electron–hole pairs. This is the *photovoltaic* effect, which is the basis of the *solar cell.* Its operation may be reversed: if, instead, current is driven through the junction by an external battery, electrons and holes recombine in the junction region, so producing photons. This is the *light-emitting diode.* If two faces of the crystal, perpendicular to the junction plane, are polished flat and made parallel to each other, the device operates as a *semiconductor laser.*

The diode

We now explain how the pn-junction acts as a diode rectifier: when a battery is connected such that the p-side is at a higher electrostatic potential than the n-side, current flows readily, but only a limited current flows when the polarity of the battery is reversed.

Suppose first that no external electrostatic potential difference is applied. In chapter 11 we explained that in an intrinsic semiconductor

Figure 12.3. The pn-junction: (*a*) the band structure at the instant the two pieces are brought together and (*b*) in dynamical equilibrium. The difference between (*a*) and (*b*) is caused by the electrostatic potential ϕ drawn in (*c*).

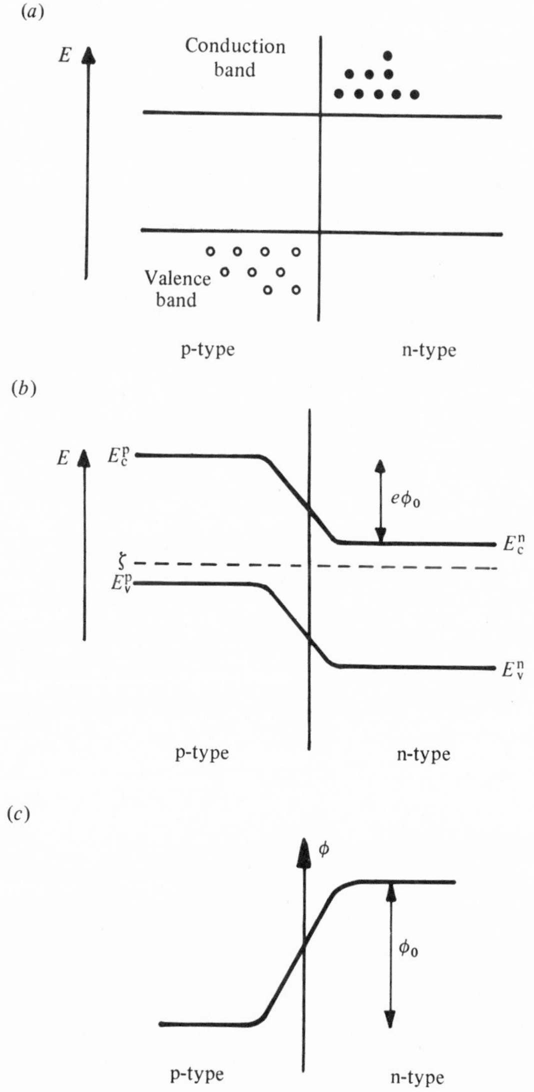

the Fermi energy ζ lies somewhere in the energy gap. Its position is shifted when the crystal is doped, and the Fermi energies of each of the two pieces, p and n, are further shifted when the pieces are pressed together to form the junction. However, statistical mechanics tells us that *any* system of electrons in dynamical equilibrium has a unique Fermi energy. So dynamical equilibrium for the pn-junction is reached when the Fermi energies ζ of the p-part and n-part coincide. We do not need to know the precise final value of the common Fermi energy, but in practice it lies somewhere in the overlap between the energy gaps on the two sides of the junction, as drawn in figure 12.3(*b*).

According to the Fermi–Dirac distribution (11.21), the number of electrons on the n-side that are above the bottom E_c^p of the conduction band on the p-side is

$$\int_{E_c^p}^{\infty} dE \frac{\rho^n(E)}{e^{(E-\zeta)/k_B T}+1} \approx \int_{E_c^p}^{\infty} dE\, \rho^n(E)\, e^{-(E-\zeta)/k_B T}, \qquad (12.7)$$

where $\rho^n(E)$ is the density of states on the n-side. Some of these electrons will be moving in the right direction to cross over to the p-side, and some of these will actually reach the p-side conduction band before collisions reduce their energy to below E_c^p. Hence there is a diffusion current of electrons from the n-side to the p-side (that is, an electric current from the p-side to the n-side):

$$I_0 \propto \int_{E_c^p}^{\infty} dE\, \rho^n(E)\, e^{-(E-\zeta)/k_B T}. \qquad (12.8)$$

Similarly, some of the electrons in the conduction band on the p-side diffuse into the n-side. In order that there be dynamical equilibrium, the corresponding current must be equal and opposite to I_0. Exactly similar considerations apply to the hole carriers.

Suppose now that an external electrostatic potential difference V is applied across the junction, by connecting the two sides of the crystal to a battery. We assume that all this potential difference appears across the junction region; if this is not a good approximation, the Ohmic potential drop IR in the rest of the crystal may be allowed for once the current I is known. With the extra driving potential V, the crystal is no longer in dynamical equilibrium; there is no longer a single Fermi energy throughout the crystal. However, on each side far from the junction there is a situation close to equilibrium, not much affected by the additional current flow caused by V. This is provided that V is not too large, so that the drift velocity superimposed on the electron motion, corresponding to the additional current, is small. Then outside the junction region each piece of crystal may, to a good approximation, be described by a Fermi–Dirac distribution, with the Fermi energy in the same place relative to the bands as it was before V was applied. (See figure 12.4, and compare it with figure 12.3(*b*) – we have chosen to define the origin $E=0$ of the energy scale such that its position relative to the n-type bands has not changed, and have drawn the figure for the case where the sign of V is such that the p-type bands move upwards in energy.) In the narrow transition region at the junction, the Fermi level is not defined: the additional electric field corresponding to V is not small, and the situation there is far from dynamical equilibrium.

Because the p-type side has the same Fermi–Dirac distribution as before (in spite of the shift of origin of E), the number of electrons in the conduction band on that side is the same as before. Thus the diffusion current arising from electrons passing from the p-side to the

Figure 12.4. (*a*) The band structure of figure 12.3(*b*) when there is an additional potential difference V whose sign is indicated in (*b*).

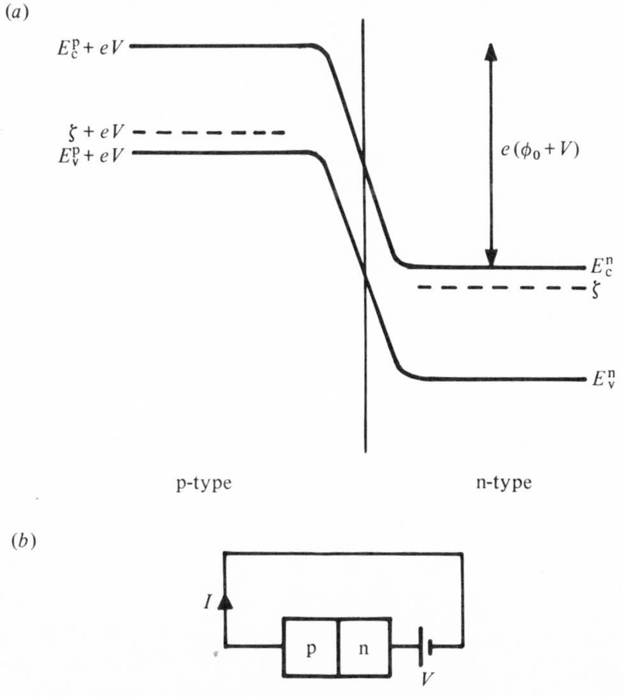

n-side is still I_0 given by (12.8). But the diffusion current in the other direction is changed. Now in the n-type region only those electrons above the level $E = E_c^p + eV$ have sufficient energy to pass over to the p-type conduction band, and so they carry a diffusion current proportional to

$$\int_{E_c^p+eV}^{\infty} dE\, \rho^n(E)\, e^{-(E-\zeta)/k_BT}. \tag{12.9}$$

Assuming that the density of states $\rho^n(E)$ varies only slowly with E, this is approximately

$$I_0\, e^{-eV/k_BT}.$$

Hence the total electric current that passes through the crystal from the n-side to the p-side is

$$I = I_0(1 - e^{-eV/k_BT}). \tag{12.10}$$

This is the contribution from electron carriers; the hole carriers contribute similarly.

I is plotted against V in figure 12.5. When V is positive, $I < I_0$, but for negative V, the current increases exponentially with the magnitude of V (until the approximations made in the derivation are no longer valid).

The junction transistor

The junction triode transistor consists of a thin slice of p-type crystal sandwiched between two pieces of n-type crystal. Alternatively it may be a pnp-sandwich, but for definiteness we consider the npn case.

Figure 12.5. Plot of current I against potential difference V for a semiconductor diode. The signs of V and I are defined in figure 12.4(b).

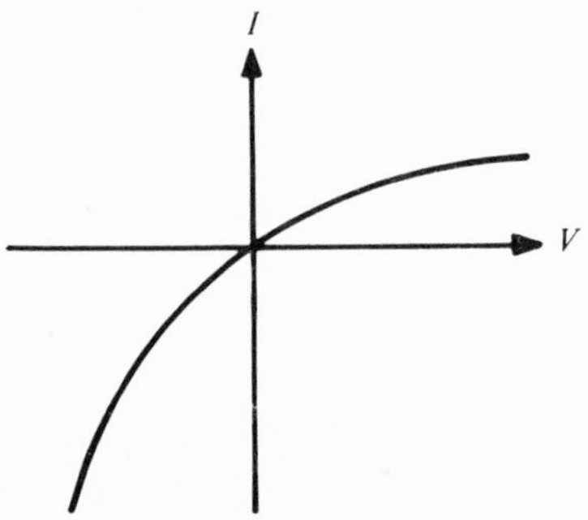

When no external potentials are applied, the disposition of the energy bands at each of the two junctions is as in figure 12.3(*b*). We draw this in figure 12.6(*a*), for the case where the two pieces of n-type material are doped with equal impurity concentrations. Suppose now that the device is connected to a pair of batteries as in figure 12.6(*b*). The p-type slice (called the *base*) is at slightly lower electrostatic potential than the left-hand n-type piece (the *emitter*), and the right-hand n-type piece (the *collector*) is at rather higher potential.

The donor impurity concentration in the n-type material is made much greater than the acceptor impurity concentration in the p-type material, so that almost all the electric current is carried by electrons rather than holes. The potential across the emitter/base circuit causes a flow of electrons through the crystal from the emitter to the base, as in the case of the diode. The base is made very thin, less than 5 μm across, so that most of these electrons diffuse across it before collisions stop or reverse their motion. The potential applied to the collector lowers the level of the bottom of the conduction band in the collector region by a large amount. Hence when the electrons arrive in the collector region from the base-region conduction band, their energy is high enough for them to be able to suffer many collisions, and cascade down the levels in the conduction band, before they are stopped. So almost all of them reach the collector terminal. That is, almost all the current of electrons that enters the emitter emerges from the collector:

$$I_c \approx I_e. \tag{12.11}$$

The emitter current I_e is controlled by the magnitude of the potential difference between the emitter and the base, as for the diode. This potential difference may be regarded as the input to the circuit; it may vary with time. The output of the circuit is the potential difference I_cR across the resistor shown in figure 12.6(*b*). This can be much greater than the input, if R is large – though R must not be too large, for then the collector potential would be lowered so much that the device would not operate. A power gain of 10^4 is not untypical.

The pnp-transistor works similarly: the signs of all the potentials and currents are reversed, and holes are the principal carriers.

By doping a single piece of host crystal, many junction transistors can be placed together on a single chip of crystal. With suitable doping techniques, such an *integrated circuit* may also contain resistors and capacitors (though not inductors), but as these cannot be

Figure 12.6. The npn-junction transistor (*a*) with no external potential, (*c*) with external potentials applied as in (*b*).

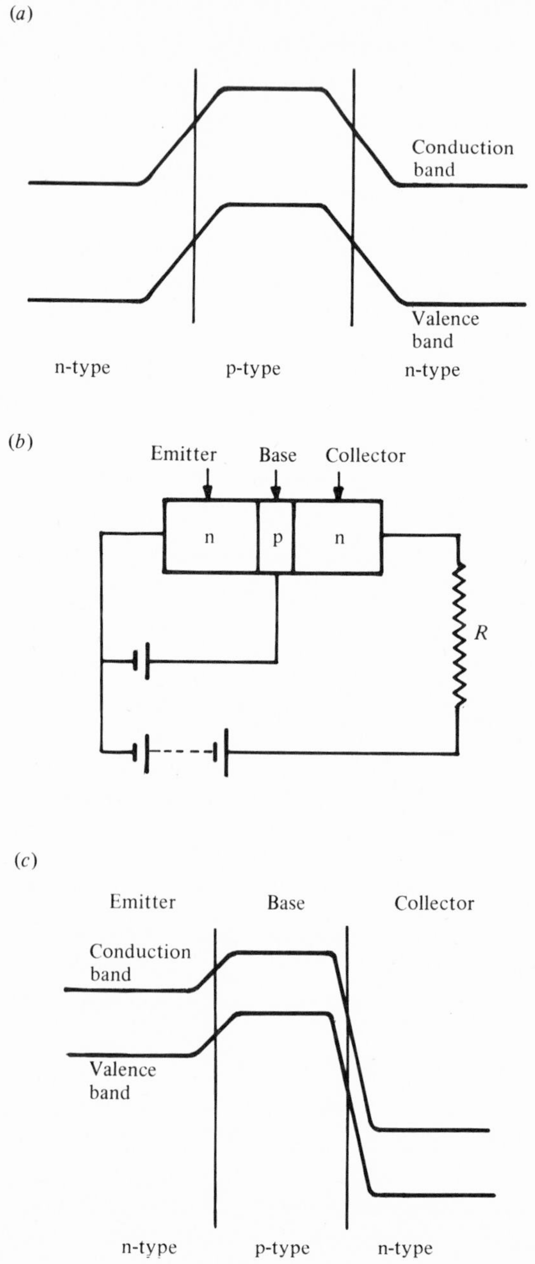

manufactured so conveniently circuits are designed in such a way as to replace them with transistors as far as possible. We have seen that the essential physics of the transistor is in the junction region, which is very thin; the remainder of the material has no active function. Hence a very large number of transistors can be accommodated in a single integrated circuit, tens of thousands or more on a crystal chip a few millimetres square.

Largely for constructional reasons, integrated circuits commonly also use a different type of transistor, the *field-effect transistor.* These can also have the advantage of much lower power consumption. There are various types of field-effect transistor; an example is the metal-oxide–silicon field-effect transistor (MOSFET), shown in figure 12.7. A wafer of insulator is sandwiched between a layer of metal, called the *gate*, and a piece of doped semiconductor. This system replaces the base of a junction transistor. If the semiconductor is p-type, it is placed between two pieces of n-type crystal; these are called the *source* and the *drain.*

The drain is raised to a positive potential relative to the source. This causes little flow of current since, as we have seen, under these conditions electrons cannot readily cross the p-type region. However, the situation changes if the gate is raised to a positive potential. Although the gate is electrically insulated from the p-type material, this causes an electric field to be set up near the upper surface of the p-type material. Consequently, the conduction and valence bands are lowered in energy in this region (figure 12.8), and the Fermi level ζ appears in the conduction band. According to the Fermi–Dirac distribution, there is then a substantial population of electrons in the conduction band in this region, even though in the bulk of the p-type material holes are the majority carriers. This *inversion layer* at the surface allows electrons to flow from the source across to the drain.

Figure 12.7. A field-effect transistor.

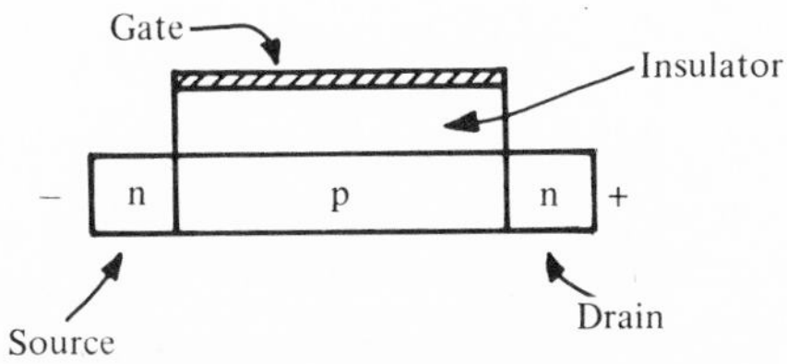

The larger the gate voltage, the greater the electron concentration in the inversion layer and the greater the current.

Two simple circuits

The simplest application of the diode is in the *demodulator* circuit, which is one of the first stages of a radio receiver. The radio-frequency carrier wave, modulated in amplitude by the lower-frequency audio signal, is applied between the p-type terminal of the diode and earth. If the other side of the diode is connected to earth through a resistor, current flows through the diode and the resistor when the input to the diode is positive. That is, the output potential across the resistor corresponds to the input with its lower half chopped off. This is shown in figure 12.9(*a*); in the circuit symbol that denotes the diode, the arrow indicates the direction in which current can pass freely. If the resistor is now replaced by a capacitor (figure 12.9(*b*)), the radio-frequency component is removed, since the impedence of the capacitor is low for high frequencies. The output is now just the audio-frequency signal, which is amplified in further stages of the radio receiver.

Notice that the rectifier is needed. To see this, suppose that the input is the very simple amplitude-modulated wave $\cos pt \cos \omega t$ where ω is the radio frequency and p is the audio frequency. But

$$\cos pt \cos \omega t = \tfrac{1}{2}[\cos(\omega + p)t + \cos(\omega - p)t]. \qquad (12.12)$$

Figure 12.8. Lowering of conduction and valence bands near the surface of the p-type material when a positive potential is applied to the gate.

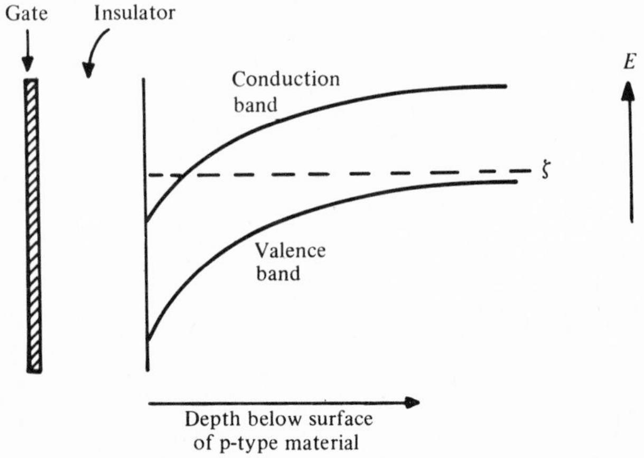

Hence, the input is the sum of two high-frequency components, and so the capacitor would short circuit the whole signal if the rectifier were not there. The signal passed by the rectifier is $F(t) = \cos pt \cos \omega t\, \theta(\cos pt \cos \omega t)$, where $\theta(x) = 1$ for $x > 0$ and $\theta(x) = 0$ for $x < 0$. If this function is represented as a Fourier integral, it is found to have a low-frequency component corresponding to the audio signal. It is not easy to calculate the Fourier transform, but a function that has much the same shape as $F(t)$ is $G(t) = \cos^2 \tfrac{1}{2}pt \cos^2 \tfrac{1}{2}\omega t$. Like $F(t)$, $G(t)$ is non-negative, and both $F(t)$ and $G(t)$ oscillate in a similar fashion. But

$$\begin{aligned} G(t) &= \tfrac{1}{4}(\cos pt + 1)(\cos \omega t + 1) \\ &= \tfrac{1}{4}(\cos pt \cos \omega t + \cos \omega t + \cos pt + 1). \end{aligned} \qquad (12.13)$$

As we have seen in (12.12), the first term in this expansion consists of high-frequency components only, as does the second term; these are short circuited by the capacitor. So the output from the circuit is $\tfrac{1}{4}(\cos pt + 1)$, which is the audio signal superimposed on a constant potential.

Figure 12.9. A radio-frequency carrier wave, amplitude modulated with an audio-frequency signal, applied to a diode circuit; (*a*) chops off the negative components of the input, (*b*) also removes the high-frequency component.

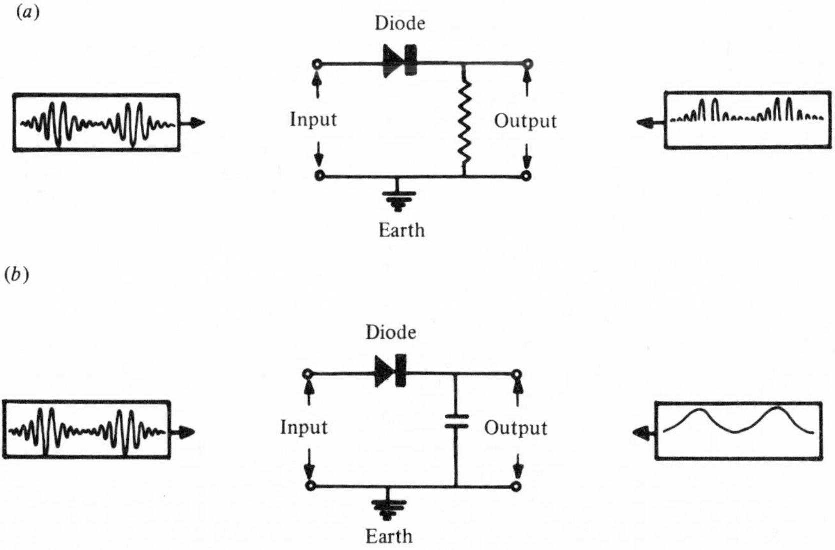

We have already described the use of the junction transistor as an amplifier. This is an example of an *analogue* circuit: the output varies continuously with the input. Many electronic instruments, in particular calculators, rather make use of *digital* circuits. In a digital circuit, a given circuit element operates in only two recognisable states, for example carrying a significant current and carrying little or no current. An example of such a circuit is the *bistable* circuit or *flip-flop*. This is drawn in figure 12.10, where two npn-junction transistors are used. Suppose that transistor 1 is conducting. Then the point A is at rather lower potential than the battery terminal E. Hence the point C, and so also the base of transistor 2, is at a comparatively low potential. Thus transistor 2 is not conducting, and so the point B is at a potential somewhat higher than point A. Hence also the point D, and so also the base of transistor 1, is at a comparatively high potential. This is as it should be if the transistor 1 is conducting, so the situation is stable. Suppose now that a negative signal is applied to the base of transistor 1, so that it stops conducting. Then the potential of point A rises, and therefore also of point C and of the base of transistor 2. Thus transistor 2 now conducts; this is the other stable situation. If either transistor 2 is switched off by a negative signal applied to its base, or transistor 1 is switched on by a positive signal applied to its base, the circuit returns to the first stable configuration.

Figure 12.10. The flip-flop circuit.

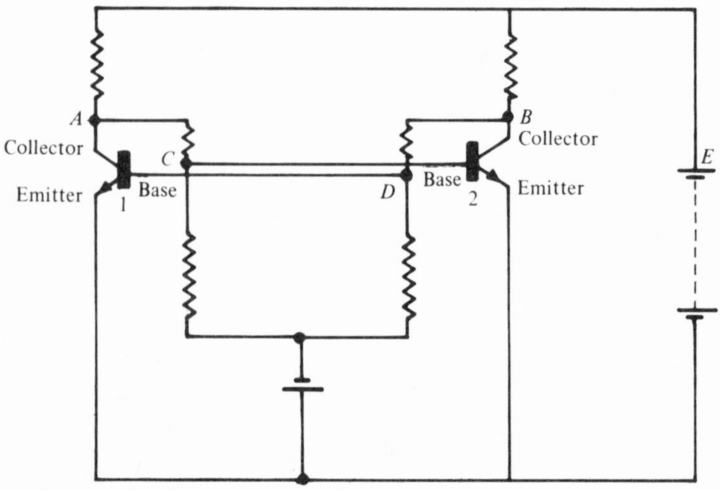

Problems

12.1 In a one-dimensional crystal an impurity atom replaces the atom at the origin. Investigate the energy-level structure, taking as a model for the potential

$$V(x) = \sum_{n\neq 0} \left[\frac{-\hbar^2}{2m} U_1 \delta(x - nL)\right] - \frac{\hbar^2}{2m} U_2 \delta(x).$$

12.2 The *tunnel diode* is a junction between p- and n-type materials which are both very heavily doped, so that the impurity levels are spread out into bands. In the n-type material the Fermi level lies in the conduction band, and in the p-type material it lies in the valence band. The transition region at the junction is made so narrow that tunnelling readily takes place across the interface.

Sketch the energy bands near the interface (a) when no external potential is applied, (b) when the n-type material is connected to a small negative potential V relative to the n-type, and (c) to a fairly large negative potential.

Sketch the graph of the current I through the junction against the potential V. (The correct curve has a region of negative 'resistance' $\mathrm{d}I/\mathrm{d}V$.)

APPENDIX A

Power-series solutions

In this appendix we describe briefly the procedure for solving differential equations by means of power-series expansions. This procedure is used for finding the stationary-state wave functions of the linear harmonic oscillator (chapter 3) and of the hydrogen atom (chapter 5).

The linear harmonic oscillator

The Schrödinger equation to be solved is given in (3.15):

$$(-\hbar^2/2m)(\mathrm{d}^2\psi/\mathrm{d}x^2)+(\tfrac{1}{2}m\omega^2x^2-E)\psi=0. \tag{A.1}$$

We are interested in solutions having the property $\psi\to 0$ as $x\to\pm\infty$, so we begin by examining the differential equation (A.1) at very large values of x. For any given eigenvalue E, when x is sufficiently large E is much less than $\frac{1}{2}m\omega^2x^2$ and so the term $-E\psi$ in the equation is then much less important than the other two terms. This suggests that it may be useful to write

$$\psi(x)=\mathrm{e}^{-m\omega x^2/2\hbar}H(x), \tag{A.2}$$

because the exponential factor has the property that, correct up to terms of leading order in x,

$$(\hbar^2/2m)(\mathrm{d}^2/\mathrm{d}x^2)\,\mathrm{e}^{-m\omega x^2/2\hbar}\sim\tfrac{1}{2}m\omega^2x^2\,\mathrm{e}^{-m\omega x^2/2\hbar}.$$

The exponential $\mathrm{e}^{+m\omega x^2/2\hbar}$ has the same property, but if we used this instead, $H(x)$ would have to obey very much more stringent conditions at $x=\pm\infty$ in order to satisfy the required boundary conditions for ψ.

Substitute (A.2) into (A.1):

$$H''-(2m\omega x/\hbar)H'+[2m(E-\tfrac{1}{2}\hbar\omega)/\hbar^2]H=0. \tag{A.3}$$

Look for a power-series solution:

$$H(x) = \sum_{s=0}^{\infty} a_s x^s. \tag{A.4}$$

We substitute this into the left-hand side of the differential equation (A.3) and equate to zero the coefficient of x^s:

$$a_{s+2} = \left[\frac{2m}{\hbar} \frac{s\omega - (E - \frac{1}{2}\hbar\omega)/\hbar}{(s+1)(s+2)}\right] a_s. \tag{A.5}$$

For any choice of a_0 and a_1, this relation determines a_s for all $s > 1$. There are two parameters a_0 and a_1 that can be chosen arbitrarily, because the differential equation (A.3) is second-order.

When s is large, (A.5) gives

$$\frac{a_{s+2}}{a_2} \sim \frac{2m\omega}{\hbar} \frac{1}{s}. \tag{A.6}$$

This means that, for any fixed value of x, the ratio of successive terms in the series (A.4) tends to zero as $s \to \infty$, and therefore the series converges. Consider now the series expansion of the function $A\,\mathrm{e}^{m\omega x^2/\hbar}\, x^B$, where A and B are constants. Its coefficients also have the property (A.6) for large s. If we make the plausible assumption, which can be justified, that the form of $H(x)$ at large x is controlled by the high powers in the series, we conclude that for large x

$$H(x) \sim A\,\mathrm{e}^{m\omega x^2/\hbar} x^B.$$

Inserting this behaviour into the expression (A.2) for ψ, we see that, whatever may be the values of A and B, we do not have the desired behaviour for ψ at infinity.

The only way to avoid this is to arrange that the series expansion of $H(x)$ terminates at some value of s, so that $H(x)$ is a polynomial. From (A.5), we see that this will happen if and only if:

either $a_1 = 0$, so that all odd powers of x are absent, and for some even value $s = 2n$, $(E - \frac{1}{2}\hbar\omega) = 2n\hbar\omega$, so that the series terminates at $s = 2n$,

or $a_0 = 0$, so that all even powers of x are absent, and for some odd value $s = (2n+1)$, $(E - \frac{1}{2}\hbar\omega) = (2n+1)\hbar\omega$, so that the series terminates at $s = (2n+1)$.

Together, these two possibilities correspond to the results quoted in (3.16) and (3.17).

The hydrogen atom

For the hydrogen atom, the differential equation to be solved is given in (5.16):

$$-\frac{\mathrm{d}^2}{\mathrm{d}r^2}[rR(r)]+\left[\frac{l(l+1)}{r^2}-\frac{Me^2}{2\pi\varepsilon_0\hbar^2 r}+\kappa^2\right]rR(r)=0. \tag{A.7}$$

We again want a solution $R(r)$ that goes to zero as $r\to\infty$.

The procedure for obtaining the solution is similar to above. However, in the previous example we knew that we had obtained both the independent solutions of the second-order differential equation (A.3) by making the expansion (A.4), because the two parameters a_0 and a_1 could be chosen freely. If we proceed here in exactly the same way, we obtain only one solution. In order to reveal both the two independent solutions, we must take a slightly more general form of series expansion; otherwise, the procedure is just the same.

When r is large, the term κ^2 in the square bracket in (A.7) dominates over the other two terms. If we neglect the other two terms, the resulting differential equation for $rR(r)$ has solutions $\mathrm{e}^{\pm\kappa r}$. By analogy with (A.2), we therefore write

$$rR(r)=\mathrm{e}^{-\kappa r}f(r). \tag{A.8}$$

Substituting this into the differential equation (A.7), we find

$$f''-2\kappa f'-\left[\frac{l(l+1)}{r^2}-\frac{Me^2}{2\pi\varepsilon_0\hbar^2 r}\right]f=0. \tag{A.9}$$

Now make the modified series expansion

$$f(r)=r^\sigma\sum_{s=0}^{\infty}a_s r^s, \tag{A.10}$$

and substitute this expansion into the left-hand side of the differential equation (A.9). The lowest resulting power of r that we find is $r^{\sigma-2}$, and on equating the coefficient of this to zero we find

$$a_0[\sigma(\sigma-1)-l(l+1)]=0. \tag{A.11}$$

This is known as the *indicial equation.* It tells us that if we want to choose the constant σ such that $a_0\neq 0$, so that the expansion (A.10) begins with the power r^σ, then either $\sigma=-l$ or $\sigma=(l+1)$.

If we make the choice $\sigma=-l$, the resulting wave function $\psi(\boldsymbol{r})$ diverges like r^{-l-1} at the origin. This means that then $\psi(\boldsymbol{r})$ does not satisfy the Schrödinger equation at $\boldsymbol{r}=0$, even though it does satisfy the equation at all other points $\boldsymbol{r}$. Because the Schrödinger equation is to be satisfied at every point $\boldsymbol{r}$, we must reject this solution.

We therefore set $\sigma = (l+1)$. We now equate to zero the coefficient of $r^{\sigma+s-1}$:

$$a_{s+1} = \frac{2\kappa(\sigma+s) - Me^2/2\pi\varepsilon_0\hbar^2}{(\sigma+s+1)(\sigma+s) - l(l+1)} a_s. \tag{A.12}$$

For any choice of a_0, this determines all the other coefficients a_s. When s is large,

$$a_{s+1}/a_s \sim 2\kappa/s. \tag{A.13}$$

Hence, arguing as before, we conclude that for large r

$$f(r) \sim A\, e^{2\kappa r} r^B,$$

which is not an acceptable behaviour. So again the series must terminate. From (A.12), we see that $[(Me^2/4\pi\varepsilon_0\kappa\hbar^2) - \sigma]$ must be a non-negative integer N, which is equivalent to the result (5.17), with $N = n - l - 1$. (Recall that we already know, from the analysis of orbital angular momentum in chapter 5, that l must be a non-negative integer.)

For the ground-state solution, we have $l = N = 0$ and $f(r) = a_0$. So the wave function is just (5.18), with the multiplicative constant chosen to normalise it correctly.

APPENDIX B

The delta function and Fourier transforms

In this appendix we present a heuristic discussion of the properties of the delta function $\delta(x)$, and we show how by using the delta function one may readily and simply understand the properties of Fourier integrals. Our presentation will be sufficient for practical purposes; a mathematically rigorous treatment may be found in pages 15 to 29 of the little book by Lighthill.†

The delta function

The delta function is an example of a *generalised function* or *distribution*. The simplest way to understand its properties is to use a limiting procedure. Define the function

$$\Delta_\varepsilon(x) = (1/\varepsilon\sqrt{\pi})\, e^{-x^2/\varepsilon^2} \qquad (\varepsilon > 0). \tag{B.1}$$

Figure B.1. Plot of the function $\Delta_\varepsilon(x)$.

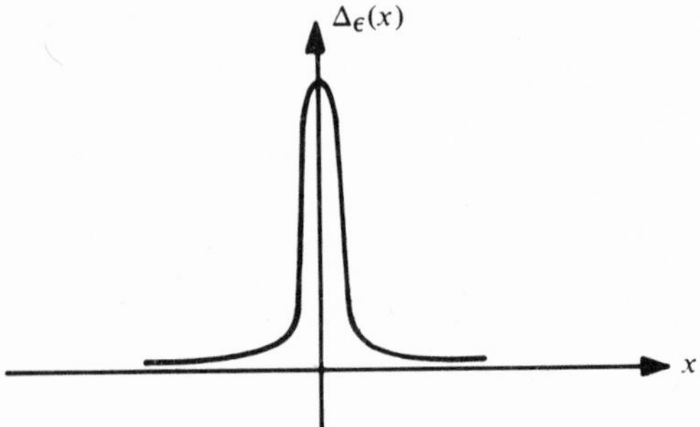

† M. J. Lighthill, *Introduction to Fourier Analysis and Generalised Functions* (Cambridge University Press, 1958).

For each value of the constant parameter ε, this function satisfies

$$\int_{-\infty}^{\infty} dx\, \Delta_\varepsilon(x) = 1. \tag{B.2}$$

When plotted against x, as in figure B.1, the function $\Delta_\varepsilon(x)$ has a peak at the origin. This peak has height $1/\varepsilon\sqrt{\pi}$ and width of order ε (exactly how we define the width does not matter), so that if ε is allowed to become very small the peak becomes very tall and very narrow. Outside the peak, the function becomes extremely small. This means that if we define an integral

$$I_\varepsilon[f] = \int_{-\infty}^{\infty} dx\, \Delta_\varepsilon(x) f(x), \tag{B.3}$$

then for a wide class of functions $f(x)$ the value of the integral when ε is very small depends almost entirely on the values that $f(x)$ takes very close to the origin, and ultimately

$$\lim_{\varepsilon \to 0} I_\varepsilon[f] = \int_{-\infty}^{\infty} dx\, \Delta_\varepsilon(x) f(0) = f(0). \tag{B.4}$$

Here we have used (B.2).

A crude definition of the delta function is

$$\delta(x) = \lim_{\varepsilon \to 0} \Delta_\varepsilon(x). \tag{B.5}$$

The mathematically rigorous definition is arrived at via (B.3) and (B.4):

$$\int_{-\infty}^{\infty} dx\, \delta(x) f(x) = f(0) \tag{B.6}$$

for a suitable class of 'test functions' $f(x)$. The test functions are ordinary functions. Different types of generalised function or distribution are defined by using different classes of test function, for example *tempered distributions* are defined using functions $f(x)$ that, together with all their derivatives, are bounded by some fixed power of x at $x = \pm\infty$.

By making a simple change of variable in (B.6), one obtains

$$\int_{-\infty}^{\infty} dx\, \delta(x-a) f(x) = f(a). \tag{B.7}$$

This may be compared with the property of the symbol δ_{ij}:

$$\sum_j \delta_{ij} f_j = f_i.$$

Hence $\delta(x-a)$ is a generalisation of δ_{ij} to the case where its discrete suffixes become continuous variables.

From the way in which we have arrived at $\delta(x)$, it follows immediately that $\delta(x)=0$ except at $x=0$. At the origin, $\delta(x)$ is positive infinite, in such a way that for any positive a and b,

$$\int_{-a}^{b} \mathrm{d}x\, \delta(x) = 1. \tag{B.8}$$

This is why it is not a function in the ordinary sense.

One may define also the derivative $\delta'(x)$ of $\delta(x)$, and higher derivatives. When ε is small, the derivative of $\Delta_\varepsilon(x)$ has two peaks close to the origin, one positive and one negative, as drawn in figure B.2. As $\varepsilon \to 0$, each of these peaks becomes very narrow and very tall, and the two peaks each approach very close to the origin. Now, an integration by parts gives

$$\int_{-\infty}^{\infty} \mathrm{d}x\, \Delta_\varepsilon'(x) f(x) = [\Delta_\varepsilon(x) f(x)]_{-\infty}^{\infty} - \int_{-\infty}^{\infty} \mathrm{d}x\, \Delta_\varepsilon(x) f(x). \tag{B.9}$$

Because of the explicit definition (B.1) of $\Delta_\varepsilon(x)$, the first term on the right-hand side vanishes unless $f(x)$ explodes violently at infinity. So, by letting $\varepsilon \to 0$, we arrive at the definition of $\delta'(x)$:

$$\int_{-\infty}^{\infty} \mathrm{d}x\, \delta'(x) f(x) = -\int_{-\infty}^{\infty} \mathrm{d}x\, \delta(x) f'(x) = -f'(0). \tag{B.10}$$

We have used (B.6) for the case where $f(x)$ is replaced by $f'(x)$. Higher derivatives of $\delta(x)$ may be defined similarly. Because the rth derivative of $\Delta_\varepsilon(x)$ changes sign r times very near the origin when ε is very small, successive derivatives of $\delta(x)$ are more and more singular at the origin.

Figure B.2. Plot of $\Delta_\varepsilon'(x)$.

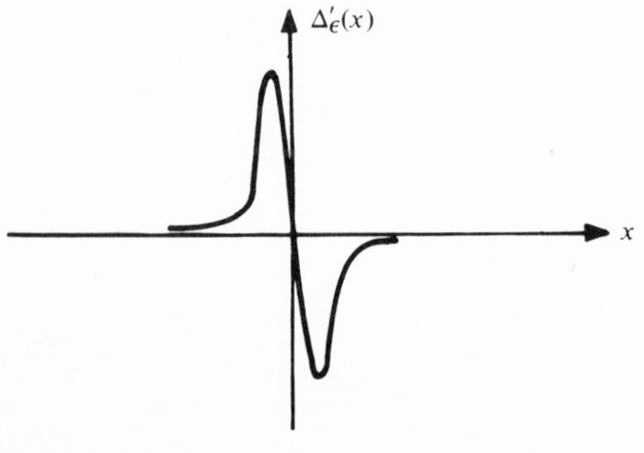

Consider now the indefinite integral of $\Delta_\varepsilon(x)$,

$$\Theta_\varepsilon(x)=\int_{-\infty}^{x} \mathrm{d}y\ \Delta_\varepsilon(y). \tag{B.11}$$

This is plotted against x in figure B.3. As $\varepsilon \to 0$, the step in this function becomes progressively steeper, until finally the function changes abruptly from 0 to 1 at the origin. In fact

$$\int_{-\infty}^{x} \mathrm{d}y\ \delta(y)=\theta(x), \tag{B.12}$$

where

$$\theta(x)=\begin{cases}1 & x>0\\ 0 & x<0.\end{cases}$$

Notice that the value of an integral such as (B.6) is unchanged if we change the value of its integrand by a finite amount at not more than a finite number of points. Hence two generalised functions that differ by a finite amount at a finite number of points are equal in the sense of distribution theory. This means, in particular, that it does not matter what value we assign to $\theta(x)$ at $x=0$; any finite value will serve, but $\theta(0)=\frac{1}{2}$ is perhaps the most natural definition.

If we differentiate (B.12) with respect to x, we obtain

$$\mathrm{d}\theta/\mathrm{d}x=\delta(x). \tag{B.13}$$

Thus distribution theory allows us to define the derivative of a function that has a finite discontinuity. Consider the function $F(x)$, defined to be equal to the ordinary differentiable function $f(x)$ for $x<a$, and

Figure B.3. Plot of $\Theta_\varepsilon(x)$.

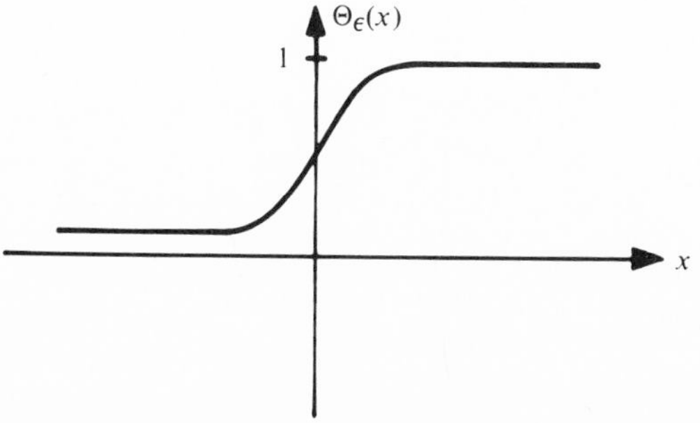

equal to the ordinary differentiable function $g(x)$ for $x > a$ (see figure B.4). We have

$$F(x) = f(x) + [g(x) - f(x)]\theta(x-a), \tag{B.14}$$

and so, applying the usual rule for differentiation of a product of functions,

$$F'(x) = f'(x) + [g'(x) - f'(x)]\theta(x-a) + [g(a) - f(a)]\delta(x-a). \tag{B.15}$$

In the last term, we obtain initially $[g(x) - f(x)]\delta(x-a)$, but we may replace x by a inside the square bracket because the delta function vanishes unless $x = a$.

Fourier transforms

In our discussion of the delta function, we used a particular function $\Delta_\varepsilon(x)$, defined by the Gaussian function (B.1), which has a peak that becomes progressively taller and narrower as $\varepsilon \to 0$. There is nothing special about this particular choice of $\Delta_\varepsilon(x)$; almost any function having this property will be just as good, provided it is chosen so as to satisfy the integral condition (B.2) for each value of ε. Another example is the function

$$\Delta_\varepsilon(x) = (1/\pi)\varepsilon/(x^2 + \varepsilon^2). \tag{B.16}$$

Consider now the integral

$$\int_{-\infty}^{\infty} dk\ e^{ikx - \varepsilon|k|}.$$

Figure B.4. Plot of $F(x)$.

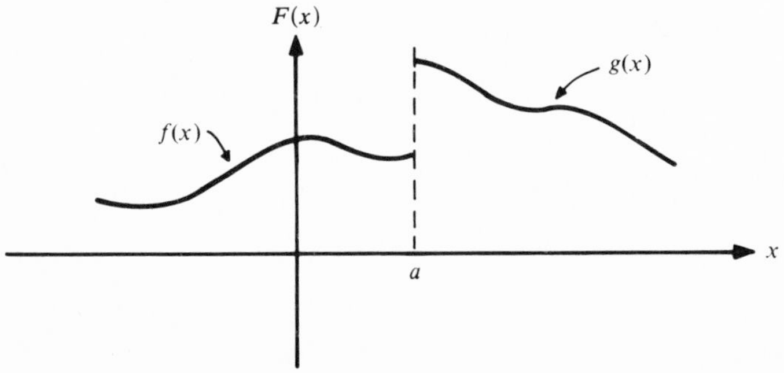

This is simple to evaluate by elementary means: it is just

$$\int_{-\infty}^{0} \mathrm{d}k\ \mathrm{e}^{ikx+\varepsilon k} + \int_{0}^{\infty} \mathrm{d}k\ \mathrm{e}^{ikx-\varepsilon k}$$

$$= 1/(\mathrm{i}x+\varepsilon) - 1/(\mathrm{i}x-\varepsilon)$$

$$= 2\pi\ \Delta_\varepsilon(x). \tag{B.17}$$

To proceed further in a rigorous fashion, we should multiply both sides by a test function $f(x)$, integrate from $-\infty$ to ∞ with respect to x, and then let $\varepsilon \to 0$. However, in the same informal sense that we understand the delta function to be defined by (B.5), we write

$$\lim_{\varepsilon\to 0} \int_{-\infty}^{\infty} \mathrm{d}k\ \mathrm{e}^{ikx-\varepsilon|k|} = 2\pi\delta(x). \tag{B.18}$$

It is usual even to write

$$\int_{-\infty}^{\infty} \mathrm{d}k\ \mathrm{e}^{ikx} = 2\pi\delta(x). \tag{B.19}$$

The integral here does not converge in the usual sense. It is *understood* that, as in (B.18), a damping factor $\mathrm{e}^{-\varepsilon|k|}$ is to be included in the integrand so as to make the integral converge, and after the integral has been calculated ε is allowed to tend to zero.

Consider now the function $\Phi(k)$, defined by the integral

$$\Phi(k) = \frac{1}{\sqrt{(2\pi)}} \int_{-\infty}^{\infty} \mathrm{d}y\ \mathrm{e}^{iky}\phi(y) \tag{B.20}$$

for some suitable function $\phi(y)$. If $\phi(y) \to 0$ sufficiently rapidly as $y \to \pm\infty$, this integral converges in the usual sense. Otherwise it is necessary to include in the integrand a damping factor $\mathrm{e}^{-\varepsilon|y|}$ and let $\varepsilon \to 0$ after the integral has been calculated. If $\phi(y)$ is bounded by some power of y as $y \to \pm\infty$, this procedure will give a convergent result. We now calculate

$$\frac{1}{\sqrt{(2\pi)}} \int_{-\infty}^{\infty} \mathrm{d}k\ \mathrm{e}^{-ikx}\Phi(k).$$

Again, a damping factor $\mathrm{e}^{-\varepsilon|k|}$ may be needed in order to make this integral converge. We insert the representation (B.20) for $\Phi(k)$ and obtain

$$\frac{1}{\sqrt{(2\pi)}}\int_{-\infty}^{\infty} dk\, e^{-ikx}\Phi(k) = \frac{1}{2\pi}\int_{-\infty}^{\infty} dk\, dy\, e^{ik(y-x)}\phi(y)$$

$$= \int_{-\infty}^{\infty} dy\, \delta(y-x)\phi(y) = \phi(x), \qquad \text{(B.21)}$$

where we have used (B.19) and then (B.7).

Putting (B.20) and (B.21) together, and changing the dummy variable y of integration in (B.20) to x, we have

$$\Phi(k) = \frac{1}{\sqrt{(2\pi)}}\int_{-\infty}^{\infty} dx\, e^{ikx}\phi(x)$$
$$\phi(x) = \frac{1}{\sqrt{(2\pi)}}\int_{-\infty}^{\infty} dk\, e^{-ikx}\Phi(k). \qquad \text{(B.22)}$$

This is the Fourier inversion theorem: if one of the relations holds, so does the other. We say that the functions Φ and ϕ are *Fourier transforms* of each other.

It may be that both the integrals in (B.22) converge without needing any additional damping factors. This is the case of *classical* Fourier transforms. Most of the textbooks consider only classical Fourier transforms; in this case it may be shown that both $\Phi(k)$ and $\phi(x)$ are ordinary functions. The possibility of including damping factors allows the very powerful extension of Fourier transform theory to include the possibility of Φ or ϕ being a distribution rather than an ordinary function. In the manipulation in (B.21) we have implicitly interchanged the order of the k- and y-integration, and when damping factors are needed we have interchanged doing the integration with taking the $\varepsilon \to 0$ limit. It is possible to show that if either Φ or ϕ, together with all its derivatives, is bounded by some power at infinity, then these interchanges are justified. Further, it may be shown that if either one of the functions Φ or ϕ has this property at infinity, then so does the other.

Notice the factor $1/\sqrt{(2\pi)}$ in front of each of the integrals in (B.22). Often, one chooses to define the Fourier transforms instead by writing $1/2\pi$ in front of one of the two integrals; then no factor appears in front of the other. However, we shall use the definition (B.22).

Simple examples of Fourier transforms

(i)
$$\delta(x) \leftrightarrow 1/\sqrt{(2\pi)}. \tag{B.23a}$$

For, because of (B.6),

$$\frac{1}{\sqrt{(2\pi)}}\int_{-\infty}^{\infty} dx\, e^{ikx}\,\delta(x) = \frac{1}{\sqrt{(2\pi)}}.$$

The inverse relation is obtained from (B.19), by replacing the dummy variable k of integration by $-k$.

(ii)
$$e^{-x^2/a^2} \leftrightarrow (a/\sqrt{2})\, e^{-a^2k^2/4}. \tag{B.23b}$$

This is an example of a classical Fourier transform: no damping factors are needed. We have

$$\frac{1}{\sqrt{(2\pi)}}\int_{-\infty}^{\infty} dx\, e^{ikx}\, e^{-x^2/a^2} = \frac{e^{-a^2k^2/4}}{\sqrt{(2\pi)}}\int_{-\infty}^{\infty} dx\, e^{-(x-\frac{1}{2}ika^2)^2/a^2}.$$

To perform the integration here, make the change of variable $(x-\frac{1}{2}ika^2) = y$ and use the standard integral

$$\int_{-\infty}^{\infty} dy\, e^{-y^2/a^2} = a\sqrt{\pi}.$$

(Strictly, after the change of integration variable the y-integration is no longer along the real axis in the complex y-plane. However, the integration may be restored to the real axis by using Cauchy's contour-integral theorem.)

(iii) For any pair of Fourier transforms $\phi(x)$, $\Phi(k)$,

$$\left(\frac{d}{dx}\right)^n \phi(x) \leftrightarrow (-ik)^n \Phi(k). \tag{B.23c}$$

This is proved immediately, by differentiating the second relation of (B.22) repeatedly with respect to x.

(iv)
$$\theta(x) \leftrightarrow \lim_{\varepsilon\to 0} \frac{i}{\sqrt{(2\pi)}}\,\frac{1}{k+i\varepsilon}, \tag{B.23d}$$

where $\theta(x)$ is the step function defined in (B.12). For, putting in the necessary damping factor,

$$\frac{1}{\sqrt{(2\pi)}}\int_{-\infty}^{\infty} dx\, e^{ikx}\,\theta(x) = \lim_{\varepsilon\to 0}\frac{1}{\sqrt{(2\pi)}}\int_{0}^{\infty} dx\, e^{ikx-\varepsilon x},$$

and this integration may be performed by elementary means. Notice that because the result is singular at $k = 0$, it is important to retain the $i\varepsilon$

in the answer. To illustrate this, we show how to verify the inverse relation. We have to evaluate

$$\lim_{\varepsilon\to 0}\frac{1}{\sqrt{(2\pi)}}\int_{-\infty}^{\infty} dk\, e^{-ikx}\left(\frac{i}{\sqrt{(2\pi)}}\frac{1}{k+i\varepsilon}\right). \qquad (B.24)$$

The integrand is analytic throughout the complex k-plane, except for a pole at $k=-i\varepsilon$. As $\varepsilon\to 0$, this pole approaches the origin, but the term $i\varepsilon$ tells us that it approaches the real axis, along which lies the contour of integration, from below rather than from above. See figure B.5. Suppose now that $x<0$. Then the exponential factor in the integral (B.24) vanishes as $k\to\infty$ in the upper half of the complex k-plane sufficiently rapidly that the integral taken over an infinite semicircle in the upper half-plane would be zero. Hence we may add this contour to the original integration contour along the real axis, without changing the value of the integral. The integration contour is now closed and, because the pole is in the lower half-plane, the integral vanishes by Cauchy's theorem. When $x>0$, we instead add on to the original integration an integration over an infinite semicircle in the lower half-plane. The integral is then evaluated by taking the residue at the pole, which now lies within the contour.

(v) Let $\phi_1(x)$ and $\phi_2(x)$ be the Fourier transforms of $\Phi_1(k)$ and $\Phi_2(k)$, respectively. Then

$$\phi_1(x)\phi_2(x)\leftrightarrow\frac{1}{\sqrt{(2\pi)}}\int_{-\infty}^{\infty} dk'\,\Phi_1(k')\Phi_2(k-k'). \qquad (B.23e)$$

Figure B.5. Integration contours for the integral (B.24).

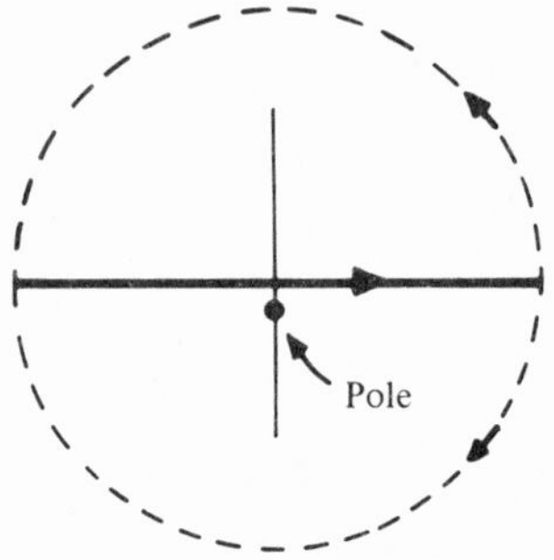

This is known as the *convolution theorem*. The proof is as follows:

$$\frac{1}{\sqrt{(2\pi)}}\int_{-\infty}^{\infty} \mathrm{d}x\, \mathrm{e}^{\mathrm{i}kx}\, \phi_1(x)\phi_2(x)$$

$$=\frac{1}{(2\pi)^{3/2}}\int_{-\infty}^{\infty} \mathrm{d}x\, \mathrm{d}k_1\, \mathrm{d}k_2\, \mathrm{e}^{\mathrm{i}(k-k_1-k_2)x}\, \Phi_1(k_1)\Phi_2(k_2)$$

$$=\frac{1}{\sqrt{(2\pi)}}\int_{-\infty}^{\infty} \mathrm{d}k_1\, \mathrm{d}k_2\, \delta(k-k_1-k_2)\Phi_1(k_1)\Phi_2(k_2)$$

and this integral is equivalent to the integral in (B.23*e*).

Finally, we prove *Parseval's theorem*. This is that, for any pair of transforms $\phi(x)$ and $\Phi(k)$,

$$\int_{-\infty}^{\infty} \mathrm{d}x\, |\phi(x)|^2 = \int_{-\infty}^{\infty} \mathrm{d}k |\Phi(k)|^2. \tag{B.25}$$

For the left-hand integral is

$$\int_{-\infty}^{\infty} \mathrm{d}x\, \phi^*(x)\phi(x)$$

$$=\frac{1}{2\pi}\int_{-\infty}^{\infty} \mathrm{d}x\, \mathrm{d}k_1\, \mathrm{d}k_2\, \mathrm{e}^{\mathrm{i}(k_2-k_1)x}\, \Phi^*(k_1)\Phi(k_2)$$

$$=\int_{-\infty}^{\infty} \mathrm{d}k_1\, \mathrm{d}k_2\, \delta(k_2-k_1)\Phi^*(k_1)\Phi(k_2).$$

APPENDIX C

Orbital-angular-momentum operators

The orbital-angular-momentum operator is defined in (5.2):

$$\hbar \boldsymbol{L} = \boldsymbol{r} \wedge (-\mathrm{i}\hbar \boldsymbol{\nabla}), \tag{C.1}$$

so that

$$\begin{aligned} L_x &= -\mathrm{i}(y\, \partial/\partial z - z\, \partial/\partial y) \\ L_y &= -\mathrm{i}(z\, \partial/\partial x - x\, \partial/\partial z) \\ L_z &= -\mathrm{i}(x\, \partial/\partial y - y\, \partial/\partial x). \end{aligned} \tag{C.2}$$

In spherical polar coordinates (see figure 5.1)

$$\begin{aligned} x &= \rho \cos \phi \\ y &= \rho \sin \phi \\ z &= \rho \cot \theta \end{aligned} \tag{C.3}$$

where $\rho = r \sin \theta$.

To calculate L_z in spherical polar coordinates, we need

$$\frac{\partial}{\partial x} = \frac{\partial \rho}{\partial x}\frac{\partial}{\partial \rho} + \frac{\partial \theta}{\partial x}\frac{\partial}{\partial \theta} + \frac{\partial \phi}{\partial x}\frac{\partial}{\partial \phi} \tag{C.4}$$

and the similar expression for $\partial/\partial y$. Here the partial derivative $\partial/\partial \rho$ is calculated keeping θ and ϕ constant, $\partial/\partial \theta$ is calculated with ρ and ϕ constant, and $\partial/\partial \phi$ is calculated with ρ and θ constant. On the other hand, each partial derivative with respect to x is calculated with y and z constant. From (C.3),

$$\begin{aligned} \mathrm{d}x &= \cos \phi\, \mathrm{d}\rho - \rho \sin \phi\, \mathrm{d}\phi \\ \mathrm{d}y &= \sin \phi\, \mathrm{d}\rho + \rho \cos \phi\, \mathrm{d}\phi \\ \mathrm{d}z &= \cot \theta\, \mathrm{d}\rho - \rho \operatorname{cosec}^2 \theta\, \mathrm{d}\theta. \end{aligned} \tag{C.5}$$

Solving these three simultaneous equations for $\mathrm{d}\rho$, $\mathrm{d}\phi$ and $\mathrm{d}\theta$ we have

$$\begin{aligned} \mathrm{d}\rho &= \cos\phi\ \mathrm{d}x + \sin\phi\ \mathrm{d}y \\ \rho\ \mathrm{d}\phi &= -\sin\phi\ \mathrm{d}x + \cos\phi\ \mathrm{d}y \\ \rho\ \mathrm{d}\theta &= \sin\theta\cos\theta\ (\cos\phi\ \mathrm{d}x + \sin\phi\ \mathrm{d}y) - \sin^2\theta\ \mathrm{d}z. \end{aligned} \tag{C.6}$$

To calculate the partial derivatives with respect to x, we set $\mathrm{d}y = \mathrm{d}z = 0$, and so have

$$\frac{\partial\rho}{\partial x} = \cos\phi, \qquad \frac{\partial\theta}{\partial x} = \frac{1}{\rho}\sin\theta\cos\theta\cos\phi, \qquad \frac{\partial\phi}{\partial x} = -\frac{1}{\rho}\sin\phi. \tag{C.7}$$

We insert these in (C.4), and calculate $\partial/\partial y$ similarly. Using these expressions, with x and y from (C.3), we calculate L_z from (C.2). The result is given in (5.5). The other two components of $\boldsymbol{L}$ may be calculated in the same way, so that altogether

$$\begin{aligned} L_x &= \mathrm{i}(\sin\phi\ \partial/\partial\theta + \cot\theta\cos\phi\ \partial/\partial\phi) \\ L_y &= -\mathrm{i}(\cos\phi\ \partial/\partial\theta - \cot\theta\sin\phi\ \partial/\partial\phi) \\ L_z &= -\mathrm{i}\ \partial/\partial\phi. \end{aligned} \tag{C.8}$$

It is then straightforward to evaluate

$$\boldsymbol{L}^2 = L_xL_x + L_yL_y + L_zL_z,$$

with the result given in (5.3).

APPENDIX D

Electrodynamics

The electric field $\mathscr{E}$ and the magnetic field $\boldsymbol{B}$ obey four equations known as *Maxwell's equations*. Maxwell wrote down these equations before either special relativity theory or quantum mechanics were known, but neither of these two developments led to any need to change the equations. They are

$$\begin{aligned} \nabla \cdot \mathscr{E} &= 0 \\ \nabla \cdot \boldsymbol{B} &= 0 \\ \nabla \wedge \mathscr{E} &= -\dot{\mathbf{B}} \\ \nabla \wedge \boldsymbol{B} &= \varepsilon_0 \mu_0 \dot{\mathscr{E}}, \end{aligned} \tag{D.1}$$

where the dots, as usual, denote partial derivatives with respect to t. We have written the equations for fields propagating in empty space. In the presence of electric charges or of a magnetically or electrically polarisable medium, the equations must be modified. The constants ε_0 and μ_0 are related to the speed of light c by

$$c^2 = 1/\varepsilon_0 \mu_0. \tag{D.2}$$

A general theorem of vector calculus allows one to deduce from the second of Maxwell's equation that, at least locally, $\boldsymbol{B}$ may be expressed in terms of a *vector potential* $\boldsymbol{A}$:

$$\boldsymbol{B} = \nabla \wedge \boldsymbol{A}. \tag{D.3}$$

By calculating $\nabla \cdot \boldsymbol{B}$, it is straightforward to verify that the second equation of (D.1) is indeed satisfied if $\boldsymbol{B}$ is expressed in this way, but the converse result is also valid.

If we insert (D.3) into the third of Maxwell's equations, we find that

$$\nabla \wedge (\mathscr{E} + \dot{\boldsymbol{A}}) = 0. \tag{D.4}$$

Another general theorem of vector calculus then allows us to deduce that $\mathscr{E}+\dot{\boldsymbol{A}}$ can be written as the gradient of a *scalar* potential ϕ:

$$\mathscr{E} = -\dot{\boldsymbol{A}} - \boldsymbol{\nabla}\phi. \tag{D.5}$$

It is again straightforward to verify that if $\mathscr{E}$ is written in this way then it automatically satisfies (D.4), and again the converse is also true.

If we take some arbitrary scalar function $f(\boldsymbol{r}, t)$ and change $\boldsymbol{A}$ and ϕ together in the following way,

$$\begin{aligned} \boldsymbol{A} &\to \boldsymbol{A} + \boldsymbol{\nabla} f \\ \phi &\to \phi - \dot{f}, \end{aligned} \tag{D.6}$$

then the fields $\boldsymbol{B}$ and $\mathscr{E}$, given by (D.3) and (D.5), remain unaltered. The fields $\boldsymbol{B}$ and $\mathscr{E}$ are physically measurable quantities, but the functions $\boldsymbol{A}$ and ϕ are not directly measurable. Hence the transformation (D.6), which is called a *gauge transformation*, can have no physical consequences. This freedom allows one to impose on ϕ and $\boldsymbol{A}$ certain constraints known as *gauge conditions*, and one example of a possible gauge condition is $\boldsymbol{\nabla}\cdot\boldsymbol{A} = 0$, which is the condition introduced in chapter 8.

In (8.4) we wrote the Hamiltonian for an electron of charge $-e$ in interaction with an electromagnetic field:

$$H = V + (\boldsymbol{p} + e\boldsymbol{A})^2/2m - e\phi. \tag{D.7}$$

In classical physics, the Hamiltonian leads to equations of motion through the elimination of $\boldsymbol{p}$ from Hamilton's equations (8.1), which are

$$\frac{\mathrm{d}\boldsymbol{r}}{\mathrm{d}t} = \frac{\partial H}{\partial \boldsymbol{p}}, \qquad \frac{\mathrm{d}\boldsymbol{p}}{\mathrm{d}t} = -\frac{\partial H}{\partial \boldsymbol{r}}. \tag{D.8}$$

Here, the operator $\partial/\partial\boldsymbol{r}$ is usually written as $\boldsymbol{\nabla}$, and $\partial/\partial\boldsymbol{p}$ denotes the corresponding differentiation with respect to $\boldsymbol{p}$. The differentiation $\partial/\partial\boldsymbol{r}$ is carried out with $\boldsymbol{p}$ kept constant, and for $\partial/\partial\boldsymbol{p}$ $\boldsymbol{r}$ is kept constant. If we apply Hamilton's equations to the Hamiltonian (D.7), we obtain

$$\begin{aligned} &\dot{\boldsymbol{r}} = (\boldsymbol{p} + e\boldsymbol{A})/m \\ &\frac{\mathrm{d}}{\mathrm{d}t}\boldsymbol{p} = -\boldsymbol{\nabla}V - \frac{e}{m}\{[(\boldsymbol{p}+e\boldsymbol{A})\cdot\boldsymbol{\nabla}]\boldsymbol{A} + (\boldsymbol{p}+e\boldsymbol{A})\wedge(\boldsymbol{\nabla}\wedge\boldsymbol{A})\} + e\boldsymbol{\nabla}\phi. \end{aligned} \tag{D.9}$$

To write the second of these equations, we have used the vector identity

$$\boldsymbol{\nabla}\boldsymbol{u}^2 = 2(\boldsymbol{u}\cdot\boldsymbol{\nabla})\boldsymbol{u} + 2\boldsymbol{u}\wedge(\boldsymbol{\nabla}\wedge\boldsymbol{u}).$$

The first equation in (D.9) gives

$$(\mathrm{d}/\mathrm{d}t)\boldsymbol{p} = m\ddot{\boldsymbol{r}} - e(\mathrm{d}/\mathrm{d}t)\boldsymbol{A}$$
$$= m\ddot{\boldsymbol{r}} - e\dot{\boldsymbol{A}} - e(\dot{\boldsymbol{r}} \cdot \boldsymbol{\nabla})\boldsymbol{A}. \qquad \text{(D.10)}$$

Here, we have used the fact that $\boldsymbol{A}$ generally varies with both t and $\boldsymbol{r}$, so that its total derivative $\mathrm{d}\boldsymbol{A}/\mathrm{d}t$ is not simply equal to $\partial\boldsymbol{A}/\partial t = \dot{\boldsymbol{A}}$, but rather $\mathrm{d}\boldsymbol{A}/\mathrm{d}t = \dot{\boldsymbol{A}} + (\dot{\boldsymbol{r}} \cdot \boldsymbol{\nabla})\boldsymbol{A}$. We reduce the right-hand side of the second equation of (D.9) using the first equation of (D.9) and also (D.3) and (D.5), and so obtain the equation of motion

$$m\ddot{\boldsymbol{r}} = -\nabla V - e\mathscr{E} - e\dot{\boldsymbol{r}} \wedge \boldsymbol{B}. \qquad \text{(D.11)}$$

APPENDIX E

Bloch waves

The Hamiltonian that describes the motion of an electron in a one-dimensional model of a crystal is unchanged when the electron coordinate is displaced through a distance L:

$$H(x+L)=H(x). \tag{E.1}$$

We define the translation operator $\hat{D}$ such that for any function $f(x)$

$$\hat{D}f(x)=f(x+L). \tag{E.2}$$

This definition applies only to functions f, not to operators such as H. However, Hf is a function, and so we have

$$\hat{D}H(x)f(x)=H(x+L)f(x+L)=H(x)\hat{D}f(x)$$

where we have used (E.1) and (E.2). Hence, for any function $f(x)$,

$$[\hat{D}, H(x)]f(x)\equiv(\hat{D}H(x)-H(x)\hat{D})f(x)=0. \tag{E.3}$$

If an operator gives zero when it is applied to any function $f(x)$, this is equivalent to saying that the operator vanishes. Thus (E.3) gives

$$[\hat{D}, H]=0. \tag{E.4}$$

As we explained at the end of chapter 4, this implies that the operators $\hat{D}$ and H have a complete set of common eigenstates; that is, the stationary states can be chosen to be eigenstates of $\hat{D}$. Let ψ be any one of these, so that

$$\hat{D}\psi(x)=c\psi(x), \tag{E.5}$$

where c is the eigenvalue of $\hat{D}$ and is a constant. Because the operator $\hat{D}$ does not correspond to an observable, it need not be Hermitian, and c can be complex.

Because of the definition (E.2) of $\hat{D}$, (E.5) gives

$$\psi(x+L)=c\psi(x). \tag{E.6}$$

The properties (E.1) and (E.6) are valid for all pairs of points $(x, x+L)$

that lie within the crystal. Suppose that the crystal extends from $x=-\infty$ to $x=+\infty$, so that (E.6) implies that

$$(\psi,\psi)\equiv\int_{-\infty}^{\infty}\mathrm{d}x\,|\psi(x)|^2=\frac{1}{|c|^2}\int_{-\infty}^{\infty}\mathrm{d}x\,|\psi(x+L)|^2. \tag{E.7}$$

By making a change of integration variable from x to $(x+L)$ we also have the identity

$$\int_{-\infty}^{\infty}\mathrm{d}x\,|\psi(x)|^2=\int_{-\infty}^{\infty}\mathrm{d}x\,|\psi(x+L)|^2. \tag{E.8}$$

Together, (E.7) and (E.8) give $|c|^2=1$, so that $c=\mathrm{e}^{\mathrm{i}\alpha}$ with α real. Without loss of generality, we may set $\alpha=kL$:

$$c=\mathrm{e}^{ikl}, \tag{E.9}$$

where k is real. According to (E.6) and (E.7), each time that we displace the coordinate x through distance L the wave function is multiplied by $\mathrm{e}^{\mathrm{i}kL}$. So for any positive or negative integer m

$$\psi(x+mL)=\mathrm{e}^{ikmL}\psi(x). \tag{E.10}$$

This relation is known as the *Bloch condition*. If we choose to write the wave function in the form (10.3), namely

$$\psi(x)=\mathrm{e}^{ikx}u_k(x), \tag{E.11}$$

it is straightforward to see that the Bloch condition requires that $u_k(x+mL)=u_k(x)$, so that $u_k(x)$ has the periodicity L of the lattice.

In order to derive the result that k is real, we needed (E.6) to be true for all values of x. This is not so when the crystal is taken to be of finite length. Then there are also solutions with k complex. In the case of such a solution, write $k=k_1+\mathrm{i}k_2$, with k_1 and k_2 real. The wave function (E.11) then contains a factor e^{-k_2x}. Depending on the sign of k_2, this is the largest at one or other of the two ends of the crystal, and it decreases exponentially with increasing distance from that end. Unless the crystal is very small, the wave function is negligibly small over most of its interior. For this reason, the solution is called a *surface state*. It turns out that surface states make a negligible contribution to the bulk properties of all but the very smallest microcrystals, and we have ignored them in this book.

HINTS FOR THE PROBLEMS

1.1 5×10^{29} photons per second.

1.2 Work in the frame in which the electron is initially at rest, so that its energy is mc^2. Let the momentum of the photon be $\boldsymbol{p}$; according to (1.4), its energy then is cp. So if the photon is absorbed, the electron afterwards has momentum $\boldsymbol{p}$ and energy $E=mc^2+cp$. This means that $E^2>m^2c^4+c^2p^2$, in violation of (1.3*b*).

1.3 The wavelength of visible light is of the order of 500 nm, so for a mass of 1 kg the answer is about 3×10^{20} years, and for an electron about 10^{-3} s.

1.4 Suppose that after the scattering the photon has momentum $\boldsymbol{p}'$. Then according to (1.4) its energy has changed from cp to cp'. Applying (1.3*b*) to the electron after the scattering gives

$$mc^2+cp-cp'=c[m^2c^2+(\boldsymbol{p}-\boldsymbol{p}')^2]^{1/2}.$$

Squaring both sides, using $\boldsymbol{p}\cdot\boldsymbol{p}'=pp'\cos\theta$, and dividing through by $2mc^3pp'/h$ gives the answer.

1.5 The calculation is similar to problem 1.4.

2.1 According to (2.7*b*), adding a constant V_0 to $V(\boldsymbol{r})$ is equivalent to instead subtracting the same constant from E. Hence, according to (2.9), the stationary-state solutions become multiplied by $e^{iV_0t/\hbar}$. This leaves unchanged the probability density $|\Psi|^2$ for all $\boldsymbol{r}$ and all t.

2.2 The allowed values of $\frac{1}{2}mv^2$ are given by (2.13*b*) so the allowed values of v are $n\hbar\pi/ma$. We need the difference between the values of v for $n=2$ and $n=1$. For the electron, this is about 3×10^6 m s^{-1}, and for the tennis ball it is about 10^{-33} m s^{-1}. In the case of the tennis ball, the answer is negligibly small: the velocity is effectively a continuous variable and quantum mechanics need not be used to describe the motion.

2.3 $C_n=(2/a)^{1/2}$ (times an arbitrary complex number of modulus 1).

$$\langle x\rangle=\int_0^a x|\psi_n(x)|^2\,dx$$

$$\langle(x-\langle x\rangle)^2\rangle=\langle x^2\rangle-\langle x\rangle^2;\quad \langle x^2\rangle=\int_0^a x^2|\psi_n(x)|^2\,dx.$$

2.4 In one dimension,

$$j(x,t)=\frac{\hbar}{2mi}\left(\Psi^*\frac{\partial\Psi}{\partial x}-\Psi\frac{\partial\Psi^*}{\partial x}\right).$$

In a stationary state, $\Psi=\psi(x)\,e^{-iEt/\hbar}$, and so $j=(\hbar/2mi)(\psi^*\psi'-\psi\psi^{*\prime})$, where $\psi'=\partial\psi/\partial x$. This is manifestly independent of t. To show that it is independent of x, calculate $\partial j/\partial x$ and use the time-independent Schrödinger equation to show that it vanishes.

In three dimensions both ρ and $\boldsymbol{j}$ are independent of t. Applying the divergence theorem to (2.17) gives $\boldsymbol{\nabla}\cdot\boldsymbol{j}=0$. This does not imply that $\boldsymbol{j}$ is independent of $\boldsymbol{r}$.

2.5 Normalise ψ such that $\int d^3\boldsymbol{r}_1\, d^3\boldsymbol{r}_2|\psi(\boldsymbol{r}_1,\boldsymbol{r}_2)|^2=1$, where the integrations over both $\boldsymbol{r}_1$ and $\boldsymbol{r}_2$ extend over all space. Then the integrand is the probability of finding the first particle in the volume element $d^3\boldsymbol{r}_1$ and the second in $d^3\boldsymbol{r}_2$. The Schrödinger equation is

$$\left[-\frac{\hbar^2}{2m_1}\nabla_1^2-\frac{\hbar^2}{2m_2}\nabla_2^2+V(\boldsymbol{r}_1,\boldsymbol{r}_2)\right]\psi=E\psi,$$

where ∇_i^2 is the ∇^2 operator constructed from $\boldsymbol{r}_i$ $(i=1,2)$.

If the particles do not interact with each other, $V=U_1(\boldsymbol{r}_1)+U_2(\boldsymbol{r}_2)$. Then there are separable solutions $\psi(\boldsymbol{r}_1,\boldsymbol{r}_2)=\psi_1(\boldsymbol{r}_1)\psi_2(\boldsymbol{r}_2)$, with

$$\left[-\frac{\hbar^2}{2m_i}\nabla_i^2+U_i(\boldsymbol{r}_i)\right]\psi_i(\boldsymbol{r}_i)=E_i\psi_i,\quad i=1,2$$

and $E=E_1+E_2$.

3.1

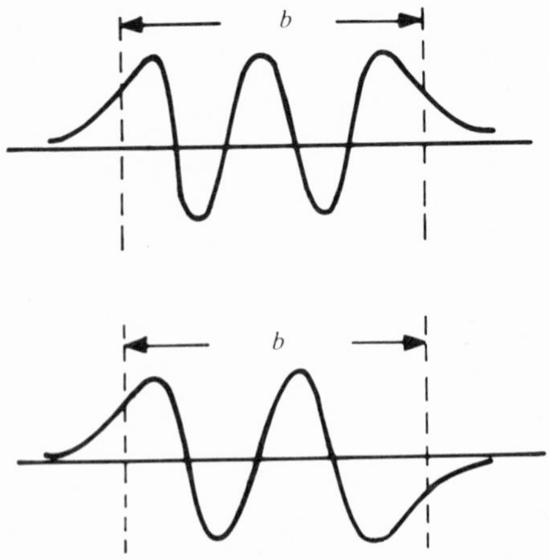

According to (3.11), within the well ψ oscillates sinusoidally, while outside the well it goes exponentially to zero. At the two edges of the well, both ψ and $d\psi/dx$ are continuous. From (3.10), the lowest levels correspond to the smallest allowed values for β and so, from (3.11*b*), their wave functions have the smallest number of oscillations.

If the values of U and b are such that in a given state the value of α is very small, the exponential fall-off of the wave function beyond the edges of the well is very slow, and then there is a good chance that the particle will be found outside the well.

3.2 With the changes of variable $x' = x - e\mathscr{E}/m\omega^2$ and $E' = E + e^2\mathscr{E}^2/2m\omega^2$, the new Schrödinger equation becomes just the same as the old one (3.15). So the new wave functions are just $\psi_n(x')$, where ψ_n is the function in (3.17), and the new energy levels are $(n+\frac{1}{2})\hbar\omega - e^2\mathscr{E}^2/2m\omega^2$.

3.3 From (3.17), the wave functions are Hermite polynomials $H_n(x)$ multiplied by an exponential factor. From appendix A, $H_0(x)$ is a constant and (see (A.5)) $H_2(x) = a_0(1 - 2m\omega x^2/\hbar)$ with a_0 a constant. To verify the orthogonality, use $\int_{-\infty}^{\infty} \mathrm{d}x\, \mathrm{e}^{-\lambda x^2} = (\pi/\lambda)^{1/2}$ and

$$\int_{-\infty}^{\infty} \mathrm{d}x\, x^2\, \mathrm{e}^{-\lambda x^2} = -\frac{\mathrm{d}}{\mathrm{d}\lambda}\int_{-\infty}^{\infty} \mathrm{d}x\, \mathrm{e}^{-\lambda x^2}.$$

3.4 Use (3.27) and the definitions $(\hbar^2k^2/2m) = E$, $(\hbar^2\kappa^2/2m) = U - E$ to show that in case (i) $|T|^2 \ll |A|^2$, and in case (ii) $|T|^2 \approx |A|^2$. Notice that (3.27) is valid also when U is negative, as in (i); then κ is imaginary.

3.5 For $x \neq 0$, the Schrödinger equation is $\psi'' = \alpha^2\psi$, where $E = -\hbar^2\alpha^2/2m$. We choose the solutions $A\,\mathrm{e}^{-\alpha x}$ in $x > 0$ and $B\,\mathrm{e}^{\alpha x}$ in $x < 0$, so that $\psi \to 0$ as $x \to \pm\infty$ as required for a bound state. This requirement also picks out the case $E < 0$, so that α is real, rather than $E > 0$. At $x = 0$, ψ must be continuous, so that $A = B$. The other condition at $x = 0$ is (3.35), with λ replaced by $-\mu$. This gives $\alpha = m\mu/\hbar^2$, and so $E = -m\mu^2/2\hbar^2$. When $U \to \infty$, (3.10) gives $\beta \to \infty$ such that $U \sim \hbar^2\beta^2/2m$. Hence if $Ub = \mu$, $\frac{1}{2}\beta b \sim m\mu/\hbar^2\beta$. Thus in (3.12*a*) the right-hand side becomes $m\mu/\hbar^2$, but in (3.12*b*) it diverges and gives no finite value for α.

3.6 The first step is achieved simply by replacing x by $-x$ throughout the Schrödinger equation (always valid for any equation!) and using $V(x) = V(-x)$. Because each value of E is supposed to correspond to only one independent wave function, it must be that $\psi(-x) = C\psi(x)$, for some constant C. Replacing x by $-x$ in this equation, we see that $\psi(x) = C\psi(-x)$. Combining these two equations gives $C^2 = 1$, so $C = \pm 1$.

4.1 $$\langle p\rangle = \int_{-\infty}^{\infty} \mathrm{d}x\, \Psi^*(-\mathrm{i}\hbar\Psi').$$

$$(\Delta p)^2 = \langle (p - \langle p\rangle)^2\rangle = \langle p^2\rangle - \langle p\rangle^2$$
$$= \int_{-\infty}^{\infty} \mathrm{d}x\, \Psi^*(-\hbar^2\Psi'') - \langle p\rangle^2.$$

4.2 $\Psi(x, t) = N_1\, \mathrm{e}^{-(x-\langle x\rangle)^2/4(\Delta x)^2}\, \mathrm{e}^{\mathrm{i}\dots}$, where $N_1 = (a^2 + 2\mathrm{i}\hbar t/m)^{-1/2}(\pi/2a^2)^{-1/4}$ and $\mathrm{e}^{\mathrm{i}\dots}$ is a phase factor that varies with both x and t.

4.3 $$\frac{\mathrm{d}}{\mathrm{d}t}\langle x\rangle = \frac{\mathrm{d}}{\mathrm{d}t}\int \mathrm{d}x\,\Psi^* x\Psi = \int \mathrm{d}x\,(\dot{\Psi}^* x\Psi + \Psi^* x\dot{\Psi}).$$

From the time-dependent Schrödinger equation, this is

$$(-\mathrm{i}\hbar/2m)\int \mathrm{d}x\,(\Psi^{*\prime\prime} x\Psi - \Psi^* x\Psi'').$$

Integrating the first term by parts twice, and using the fact that for a wave packet the wave function Ψ vanishes at $x = \pm\infty$, gives the result. The second part is done similarly.

4.4 $\langle E\rangle = \langle p^2\rangle/2m + \frac{1}{2}m\omega^2\langle x^2\rangle$.

Because $\langle p\rangle$ and $\langle x\rangle$ are zero, this is

$(\Delta p)^2/2m + \frac{1}{2}m\omega^2(\Delta x)^2$.

So, from the uncertainty relation (4.13),

$\langle E\rangle \geqslant \hbar^2/8m(\Delta x)^2 + \frac{1}{2}m\omega^2(\Delta x)^2$.

The minimum value of the right-hand side occurs when $(\Delta x)^2 = \hbar/2m\omega$, giving $\langle E\rangle \geqslant \frac{1}{2}\hbar\omega$.

4.5 The maximum accuracy is achieved by attempting to drop the dart from directly above the target point, but the uncertainty principle implies that in practice the point of release is displaced through a horizontal distance Δx and the dart has a horizontal component of momentum Δp (in the same direction) at the time of release, where $\Delta x\ \Delta p \geqslant \frac{1}{2}\hbar$. Since the time to drop a height l is $(2l/g)^{1/2}$, the dart hits the ground at distance

$R = \Delta x + (2l/g)^{1/2}\,\Delta p/m \geqslant \Delta x + (2l/g)^{1/2}\hbar/2m\ \Delta x.$

The minimum occurs for $\Delta x = (l\hbar^2/2m^2 g)^{1/4}$, giving $R \geqslant (8l\hbar^2/m^2 g)^{1/4}$.

4.6 The identity (4.16) is proved by integration by parts.

4.7 $$\int \mathrm{d}^3\boldsymbol{r}\,\Psi^*\left(-\frac{\hbar^2}{2m}\nabla^2\Psi\right) = -\frac{\hbar^2}{2m}\int \mathrm{d}^3\boldsymbol{r}\,\boldsymbol{\nabla}\cdot(\Psi^*\boldsymbol{\nabla}\Psi) + \frac{\hbar^2}{2m}\int \mathrm{d}^3\boldsymbol{r}\,\boldsymbol{\nabla}\Psi^*\cdot\boldsymbol{\nabla}\Psi.$$

The divergence theorem transforms the first integral to an integral over a surface that encloses the system so that Ψ vanishes over this surface; this integral therefore vanishes. The second integral is $(\hbar^2/2m)\int \mathrm{d}^3\boldsymbol{r}|\boldsymbol{\nabla}\Psi|^2$, which is positive. It follows that in each bound state

$$E = \langle H\rangle > \langle V\rangle = \int \mathrm{d}^3\boldsymbol{r}\ V|\Psi|^2 \geqslant \int \mathrm{d}^3\boldsymbol{r}\ V_0|\Psi|^2 = V_0,$$

where V_0 is the minimum value of V.

4.8 This follows immediately from (4.16) and the fact that q must be real, because $\hat{Q}$ is Hermitian.

4.9 By taking linear combinations of the given wave functions, and applying $\hat{Q}$ to them, one finds that the complete set of eigenfunctions is

$\phi_1 = (\psi_1 + \psi_2)/\sqrt{2},\ \phi_2 = (\psi_1 - \psi_2)/\sqrt{2},\ \psi_3, \psi_4, \ldots, \psi_n.$

The first of these has eigenvalue $+1$, so the system is left in this state after the first measurement. At a later time t the wave function is

$\Psi = (\psi_1\,\mathrm{e}^{-\mathrm{i}E_1 t/\hbar} + \psi_2\,\mathrm{e}^{-\mathrm{i}E_2 t/\hbar})/\sqrt{2}.$

The required probability is

$$|(\Psi, \phi_1)|^2 = \tfrac{1}{4}|e^{iE_1t/\hbar} + e^{iE_2t/\hbar}|^2 = \tfrac{1}{2}\{1 + \cos[(E_1 - E_2)t/\hbar]\}.$$

4.10 This follows immediately from the time-dependent Schrödinger equation.

4.11 (i) is just the definition (4.16); (ii), (iii) and (iv) follow immediately from the definition of the † symbol given in the question.

The expectation value in the state Ψ is $(\Psi, \hat{F}^\dagger\hat{F}\Psi)$, where $\hat{F} = \hat{Q} - i\lambda\hat{R}$ and so, because $\hat{Q}$ and $\hat{R}$ correspond to observables and therefore are Hermitian, $\hat{F}^\dagger = \hat{Q} + i\lambda\hat{R}$. From (iv), $(\Psi, \hat{F}^\dagger\hat{F}\Psi) = (\Phi, \Phi)$, which from its definition $\int d^3\boldsymbol{r}|\Phi(\boldsymbol{r}, t)|^2$ is real and $\geqslant 0$. Multiplying out the terms in $\hat{F}^\dagger\hat{F}$,

$$(\Phi, \Phi) = \langle\hat{Q}^2\rangle - i\lambda\langle[\hat{Q}, \hat{R}]\rangle - \lambda^2\langle\hat{R}^2\rangle.$$

This expression has its minimum when $\lambda = -\tfrac{1}{2}i\langle[\hat{Q}, \hat{R}]\rangle/\langle\hat{R}^2\rangle$, whence the required inequality follows. The uncertainty principle (4.27) is obtained from this inequality by making the replacements $\hat{Q} \to \hat{Q} - \langle\hat{Q}\rangle$, $\hat{R} \to \hat{R} - \langle\hat{R}\rangle$.

5.1 Under the rotation, $\theta \to \theta$ and $\phi \to \phi - \alpha$. So, according to (5.7), Y_{lm} is multiplied by $e^{-im\alpha}$.

5.2 $e^2/4\pi\varepsilon_0 r^2$ is a force, so $\dim(e^2/4\pi\varepsilon_0) = ML^3T^{-2}$. The momentum operator is $(-i\hbar\boldsymbol{\nabla})$, so $\dim(\hbar) = ML^2T^{-1}$. $a_0 = (4\pi\varepsilon_0\hbar c/e^2)(\hbar/mc) \approx 5\times10^{-11}$ m.

5.3 The force towards the centre of the orbit is $mv^2/a = e^2/4\pi\varepsilon_0 a^2$. So $\hbar L = mva = (mae^2/4\pi\varepsilon_0)^{1/2}, = n\hbar$ say. The corresponding energy is

$$\tfrac{1}{2}mv^2 - e^2/4\pi\varepsilon_0 a = -e^2/8\pi\varepsilon_0 a = me^4/32\pi^2\varepsilon_0^2\hbar^2 n^2$$

as in (5.17). When $n = 1$, the velocity is $e^2/4\pi\varepsilon_0\hbar = c/137$, so that $v \ll c$ and non-relativistic kinematics are justified.

5.4 When $l = 0$, the wave function depends on r only. According to (5.13), it satisfies

$$[rR(r)]'' + \frac{2mE}{\hbar^2}[rR(r)] = -U\delta(r-a)[rR(r)].$$

For $r \neq a$, the solutions to this equation are $e^{\pm\kappa r}$ where $\kappa = (2mE/\hbar)^{1/2}$. Since we want a bound-state solution, where $R \to 0$ as $r \to \infty$, and since, for the reason explained in the text, we require R to be non-infinite at $r = 0$, we take

$$rR = \begin{cases} A\,e^{-\kappa r} & r > a \\ B(e^{-\kappa r} - e^{\kappa r}) & r < a. \end{cases}$$

At $r = a$, R must be continuous, so that $A\,e^{-\kappa a} = B(e^{-\kappa a} - e^{\kappa a})$. Also, from the differential equation, $\text{disc}\,[rR(r)]'|_{r=a} = -U[rR(r)]|_{r=a}$ (compare with (3.35)). So

$$-\kappa A\,e^{-\kappa a} + \kappa B(e^{-\kappa a} + e^{\kappa a}) = -UA\,e^{-\kappa a}.$$

Eliminating A and B gives the equation that determines the energies of the bound states: $\kappa(\coth\kappa a - 1) = U$.

5.5 An extra term $-me^2b/2\pi\varepsilon_0\hbar^2r^2$ appears in the square brackets in (A.7) and (A.9); the substitution (A.8) is the same. The square bracket in the indicial equation (A.11) acquires the extra term $-me^2b/2\pi\varepsilon_0\hbar^2 = -2b/a_0$, where a_0 is the Bohr radius given in (5.18).

The required root of the indicial equation is $\sigma = l+1-D(l)$. The denominator of the right-hand side of (A.12) has an extra term, but the numerator is still the same and so again $me^2/4\pi\varepsilon_0\kappa\hbar^2-\sigma$ must be an integer $N = n-l-1$.

5.6 By analogy with (5.20), write $r = x_1 - x_2$ and $R = \frac{1}{2}(x_1+x_2)$. Then, with $\psi(x_1, x_2) = \psi_1(r)\psi_2(R)$, when the centre of mass is at rest we have

$$\left(-\frac{\hbar^2}{m}\frac{d^2}{dr^2}+\frac{1}{2}m\omega^2r^2\right)\psi_1(r) = E\psi_1(r)$$

and $\psi_2''(R) = 0$ (see (5.23)). But the first of these equations is obtained from the equation, (3.15), for the linear harmonic oscillator by making the changes $m \to \frac{1}{2}m$ and $\omega \to \sqrt{2}\omega$. So the energy eigenvalues are $E = (n+\frac{1}{2})\hbar\omega\sqrt{2}$.

6.1 For $x \neq \pm R$, the Schrödinger equation is $\psi'' = \kappa^2\psi$, where $E = -\hbar^2\kappa^2/2m$. The potential is symmetric about $x = 0$ and so (see problem 3.6) the wave functions have definite parity: either

$$\psi(-x) = \psi(x) \quad \text{or} \quad \psi(-x) = -\psi(x).$$

With such wave functions, imposing the appropriate continuity and discontinuity conditions at $x = \pm R$ guarantees that they are satisfied also at $x = -R$. At $x = \pm R$, ψ is continuous and disc $\psi' = -U\psi$. Hence $B\,e^{-\kappa R} = A(e^{\kappa R} \pm e^{-\kappa R})$ and

$$-\kappa B\,e^{-\kappa R} - \kappa A(e^{\kappa R} \mp e^{-\kappa R}) = -UB\,e^{-\kappa R}.$$

Eliminating A and B gives $U - \kappa = \kappa \tanh \kappa R$ or $\kappa \coth \kappa R$.

6.2 The equilibrium value of R corresponds to the minimum of the integral $V(R)$ given in (6.12). According to (6.1), the only length scale in the problem is that set by a_0, so the required value of R is proportional to a_0. From (5.18), the muon-bound molecule is therefore some 200 times smaller.

6.3 As in (6.10*a*), for large R we have $H_{11} \approx E_{01}$ and $H_{22} \approx E_{02}$. Eliminate c_1 and c_2 between the two equations (6.4). The resulting equation for E has one solution near E_1 and the other near E_2. To examine the solution near E_1, write $E = E_1 + \delta$ where δ is assumed small, and insert this in the equation. This gives that $(E_2 - E_1)\delta$ is quadratic in small quantities.

6.4 With the obvious notation, the symmetry of the problem results in $H_{11} = H_{22} = H_{33}$ and $H_{12} = H_{21} = H_{23} = H_{32} = H_{13} = H_{31}$. Write $\phi = c_1\phi_1 + c_2\phi_2 + c_3\phi_3$ and obtain the equations corresponding to (6.4). Eliminate c_1, c_2 and c_3 by making the determinant of their coefficients vanish. The resulting equation for E has two degenerate solutions $E = (H_{11} - H_{12})/(1 - K_{12})$ and a third solution $E = (H_{11} + 2H_{12})/(1 + 2K_{12})$.

7.1 Take the z-axis in the direction of $\mathscr{E}$. In (7.1), take H_0 as the Hamiltonian before the field is introduced and $H_1 = e\mathscr{E}z$. With the ground-state wave function (6.1), the energy shift (7.7) is

$$\Delta E = (\pi a_0^3)^{-1} \int d^3\boldsymbol{r}\, e\mathscr{E}z\, e^{-2r/a_0}$$

$$= (2/a_0^3) \int_0^\infty dr\, r^2\, e^{-2r/a_0} \int_{-1}^{1} d(\cos\theta) e\mathscr{E}r \cos\theta.$$

The $\cos\theta$ integration gives zero, so there is no first-order energy shift.

7.2 From appendix A, $H_0(x)$ is a constant, so the normalised unperturbed ground-state wave function is $\psi_0(x) = (\pi\hbar/m\omega)^{-1/4}\, e^{-m\omega x^2/2\hbar}$. The energy shift to first order is $\Delta E = \int dx\, \varepsilon x^2 \psi_0^2 = \varepsilon\hbar/2m\omega$. The perturbed Hamiltonian is $p^2/2m + \frac{1}{2}m\omega'^2 x^2$, where $\omega'^2 = \omega^2 + 2\varepsilon/m$. The exact value of the new ground-state energy is $\frac{1}{2}\hbar\omega' = \frac{1}{2}\hbar\omega + \varepsilon\hbar/2m\omega$ + higher-order terms.

7.3 (i) Let ψ be an eigenfunction of the Hamiltonian $H_0 + H_1$, and write it as the linear combination $c_1\phi_1 + c_2\phi_2$. From the Schrödinger equation, $H\psi = E\psi$,

$$c_1 H_1 \phi_1 + c_2 H_1 \phi_2 = c_1(E - E_1)\phi_1 + c_2(E - E_2)\phi_2.$$

To this equation, apply in turn the operations $\int d^3\boldsymbol{r}\, \phi_1^*$ and $\int d^3\boldsymbol{r}\, \phi_2^*$, using the orthonormality of ϕ_1 and ϕ_2 (see (4.18)). Eliminating c_1 and c_2 then gives

$$\frac{(\phi_1, H_1\phi_1) + E_1 - E}{(\phi_2, H_1\phi_1)} = \frac{(\phi_1, H_1\phi_2)}{(\phi_2, H_1\phi_2) + E_2 - E}.$$

This is a quadratic equation for E, with two solutions. Expanding the square root in the solutions gives, to first order, the same result as perturbation theory.

7.4 According to (3.4) and (3.5), the state of energy $3\hbar^2\pi^2/2ma^2$ has wave function

$$\psi_{111} = (8/a^3)^{1/2} \sin \pi x/a \sin \pi y/a \sin \pi z/a.$$

The perturbation $-e\mathscr{E}(t)x$ can induce a transition to the state whose wave function is

$$\psi_{211} = (8/a^3)^{1/2} \sin 2\pi x/a \sin \pi y/a \sin \pi z/a,$$

but not to either of the other two states of energy $3\hbar^2\pi^2/ma^2$. The transition probability is calculated from (7.22), with

$$I(\omega) = \int_{-\infty}^{\infty} dt'\, e^{i\omega t' - t'^2/T^2}$$

(see (B.23*b*)).

7.5 The normalised wave functions are $\psi_0(x) = (a/\pi)^{1/4}\, e^{-\frac{1}{2}ax^2}$ and $\psi_1(x) = (2\pi)^{1/2}(a/\pi)^{3/4} x\, e^{-\frac{1}{2}ax^2}$, where $a = m\omega/\hbar$. The required probability is

$$\frac{1}{\hbar^2}\left| \int_{-\infty}^{\infty} dt\, dx\, \psi_0(x)\psi_1(x)\lambda x \delta(x - vt)\, e^{i\omega t} \right|^2$$

$$= (2a^2 v^2 \lambda^2/\hbar^2) \left| \int_{-\infty}^{\infty} dt\, t\, e^{-a^2 v^2 t^2 + i\omega t} \right|^2.$$

The integral is

$$-\mathrm{i}\frac{\partial}{\partial\omega}\int_{-\infty}^{\infty}\mathrm{d}t\,\mathrm{e}^{-a^2v^2t^2+\mathrm{i}\omega t}=-\mathrm{i}\frac{\partial}{\partial\omega}[(\pi^{1/2}/av)\,\mathrm{e}^{-\omega^2/4a^4v^4}]$$

(see (B.23*b*)).

7.6 The probability is given by (7.31), with $\phi_i=(2/a)^{1/2}\sin\pi x/a$ and $\phi'_k=(4/a)^{1/2}\sin 2\pi x/a$. Note that the integration is from 0 to $\frac{1}{2}a$, not up to a. The answer is $32/9\pi^2$.

8.1 Equation (5.17) gives $\hbar\omega=3me^4/32\pi^2\hbar^3\varepsilon_0^2$. For a photon, $K=\omega/c$. So, with (5.18), $Ka_0=3e^2/8\pi\varepsilon_0\hbar c=(3/2)(1/137)$. Hence if it is not the case that $Kr\ll 1$, then $r/a_0\gg 1$ and the exponential in the wave function (5.18) is very small.

8.2 From (8.6), the perturbation is $H_1(e/M)\boldsymbol{A}\cdot\hat{\boldsymbol{p}}$. With $\boldsymbol{A}=\frac{1}{2}\boldsymbol{r}\wedge\boldsymbol{B}$ this is $-(e/2M)\boldsymbol{B}\cdot\boldsymbol{r}\wedge\hat{\boldsymbol{p}}=-(e/2M)\boldsymbol{B}\cdot\boldsymbol{L}$. The common eigenstates are described by the wave functions (5.12); this would be true not just for the hydrogen atom, but for any system for which the Hamiltonian H_0 was spherically symmetric before the magnetic field was applied. We explained after (5.13) that the wave functions (5.12) have the property that $(\psi, H_0\psi)$ is independent of m. On the other hand, $(\psi, H\psi)=(\psi, H_0\psi)-(e/2M)Bm$.

9.1 The stationary states $\psi_\pm$ are not eigenstates of the electric-dipole-moment operator $\hat{\boldsymbol{\mu}}$, but we may calculate the expectation of $\hat{\boldsymbol{\mu}}$ in these states. Because the eigenstates ϕ_1 and ϕ_2 correspond to different eigenvalues $\pm\mu_0$, they are orthogonal. Hence, using (9.2),

$$(\psi_+,\hat{\boldsymbol{\mu}}\psi_+)=\tfrac{1}{2}(\phi_1,\hat{\boldsymbol{\mu}}\phi_1)+\tfrac{1}{2}(\phi_2,\hat{\boldsymbol{\mu}}\phi_2)=0,$$

and similarly for the ψ_--state.

9.2 The answer is 4.6×10^9 K.

9.3 The uncertainty relation (4.13) implies that when the width of the beam is known to be Δ, the photons can have a transverse momentum component of at least $\hbar/2\Delta$. The total momentum of each photon is calculated from the de Broglie relation (1.5), and is $2\pi\hbar/\lambda$. So $\theta\geqslant(\hbar/2\Delta)(2\pi\hbar/\lambda)=\lambda/4\pi\Delta$ radians.

In the case of the torch, $\theta=\frac{1}{2}w/f$, where w is the width of the filament and f is the focal length of the reflecting mirror. With $f=10$ mm, w has to be 10^{-6} m to achieve the same value of θ. It would not be possible to construct a filament of this size with a reasonable light output.

9.4 The average number of particles in the given state is $n_{qrs}=C\,\mathrm{e}^{-E_{qrs}/k_BT}$, where $C=N(\sum_{Q,R,S}\mathrm{e}^{-E_{QRS}/k_BT})^{-1}$ and from (3.5), $E_{qrs}=(\hbar^2\pi^2/2mL^2)(q^2+r^2+s^2)$. (The fact that the states in this problem are degenerate makes no difference here; the degeneracy enters only if we calculate the number of particles in a level rather than in a state, as in (9.10).) The average fraction of particles that are in the state is n_{qrs}/N, which is also the probability that a given particle is in the state.

The velocity of the particle is $\boldsymbol{v} = \boldsymbol{p}/m = \hbar\boldsymbol{k}/m = (\hbar\pi/mL)(q, r, s)$. As $L \to \infty$, the allowed values of E_{QRS} become arbitrarily close, and

$$C \sim N\left[(mL/\hbar\pi)^3 \int d^3\boldsymbol{v}\; e^{-\frac{1}{2}mv^2/k_BT}\right]^{-1}$$

$$= N[(mL/\hbar\pi)^3(2\pi k_BT/m)^{3/2}]^{-1}.$$

In this limit, the number of allowed sets of values of (q, r, s) corresponding to the volume element $d^3\boldsymbol{v}$ of velocity space $\sim (mL/\hbar\pi)^3\, d^3\boldsymbol{v}$. If $d^3\boldsymbol{v}$ is infinitesimal, so that all the corresponding states have essentially the same energy, the average number of particles in these states is $C(mL/\hbar\pi)^3\, d^3\boldsymbol{v}\; e^{-\frac{1}{2}mv^2/k_BT}$. Integrate over v_y and v_z to obtain the number having v_x in the desired interval.

Notice that the final answer is in fact valid for a large box of any shape.

9.5 Take the x-direction normal to the membrane, and use the result of the previous problem. A particle whose x-component of velocity is u will reach the wall containing the membrane during the time interval Δt if it is within a distance $u\,\Delta t$ of the wall, and a fraction A/L^2 of such particles will hit the membrane region of the wall. Hence the number of particles in time Δt that pass through the membrane is

$$N\int_{u_0}^{\infty} du\; (m/2\pi k_BT)^{1/2}\, e^{-mu^2/2k_BT}(u\,\Delta t/L)(A/L^2).$$

9.6 At $T = 0$, $n_i = 0$ for $E_i > \zeta$ and $n_i = g_i$ for $E_i < \zeta$. This is because at $T = 0$ the energy of the system is minimised, but the exclusion principle does not allow more than one electron per state.

10.1 The Bloch waves here have $u_k(x)$ in (10.3) equal to a multiple of $e^{i2r\pi x/L}$, where r is any integer, and $V + E = \hbar^2(k + 2r\pi/L)^2/2m$. The dispersion curves are shown in the figure on page 170. In this model, there are no gaps between the bands, but these appear when the model is modified so as to become the nearly-free-electron model.

10.2 The three components of $\boldsymbol{k}$ satisfy $k_i = 2n_i\pi/V^{1/3}$, where the n_i are integers. Hence there are $V/(2\pi)^3$ allowed vectors per unit volume of $\boldsymbol{k}$-space. Taking account of the two spin states of each electron, the number of electron states having $|\boldsymbol{k}| \leq k_F$ is therefore $(\frac{4}{3}\pi k_F^3)V/4\pi^3$. At $T = 0$, the NZ electrons occupy all the states up to $k_F = (3\pi^2NZ/V)^{1/3}$.

10.3 (i) For $0 < x < L$, the potential is zero and so ψ and u_k have the forms shown in the second equations of (10.4) and (10.5), respectively. For $-L < x < 0$, u_k is obtained from the second equation of (10.5) by replacing x by $(x + L)$. Impose the conditions that, at $x = 0$, ψ is continuous and disc $\psi' = -U\psi(0)$ – see (3.35). This gives

$$\cosh \alpha L - (U/2\alpha)\sinh \alpha L = \cos kL.$$

(ii) In $0 < x < L$, take ψ and u_k as before, and in $L < x < 2L$ take similar forms but with different coefficients. Impose the conditions on ψ and disc ψ' at $x = L$. Because $V(x)$ now has period $2L$, so does $u_k(x)$. Use this to obtain $u_k(x)$ in $-L<x<0$ from $u_k(x)$ in $L<x<2L$, and impose the conditions on ψ and disc ψ' at $x=0$. Eliminate the four coefficients from the four conditions.

10.4 $H=\hat{p}^2/2m+\sum_r V_0(x-rL)$,

where

$E=(\psi, H\psi)/(\psi, \psi)$

and where

$$(\psi, \psi)=\sum_{m,n} e^{ik(m-n)L}\int dx\, \phi^*(x-nL)\phi(x-mL),$$

$$(\psi, H\psi)=\sum_{m,n} e^{ik(m-n)L}\int dx\, \phi^*(x-nL)H\phi(x-mL).$$

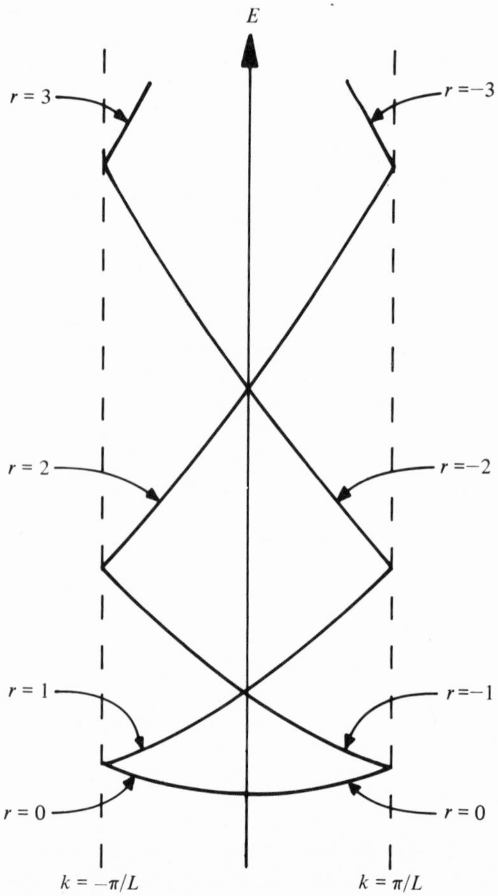

But from the definition of ϕ,

$$H\phi(x-mL)=E_0\phi(x-mL)+\sum_{r\neq m} V_0(x-rL)\phi(x-mL).$$

The first term in $(\psi, H\psi)$ contributes E_0 to E, and the second contributes

$$\frac{\sum_{n,r\neq m} e^{ik(m-n)L}\int dx\,\phi^*(x-nL)V_0(x-rL)\phi(x-mL)}{\sum_{m,n} e^{ik(m-n)L}\int dx\,\phi^*(x-nL)\phi(x-mL)}.$$

Because the binding is tight and the atoms are not very close together, the main contribution to the denominator comes from those terms having $m=n$, so that the denominator is approximately equal to N, the number of atoms. Likewise, the main contribution to the numerator comes from those terms having $r=n=m\pm 1$. There are N pairs of such terms and together they give the desired result.

10.5 In the case of periodic boundary conditions, the allowed values of k are given by (10.13), and are used to form Bloch waves $\psi_k(x)=e^{ikx}u_k(x)$. In the other case, take the ends of the crystal to be $x=0$ and $x=NL$. Just as for the free particle in a box the wave functions are sine functions, so here they will be combinations of ψ_k and ψ_{-k}. Define the normalisation of u_k and u_{-k} such that they are equal to 1 at $x=0$; because of their periodicity they are then also equal to 1 at $x=NL$. The required wave functions are then $\psi_k(x)-\psi_{-k}(x)$; this combination is chosen so as automatically to vanish at $x=0$, and it will also vanish at $x=NL$ if $k=n\pi/NL$ with $n=1, 2, 3, \ldots$. We here restrict k to positive values, because the corresponding negative values would again give the same wave functions, apart from a sign, and so do not correspond to additional quantum-mechanical states. In the interval (10.14), the number of allowed values of k is thus the same as in (10.13).

11.1 (i) From (10.13), there is one allowed wave vector in the interval $2\pi/NL$ of k-space. Taking into account the two spin states, there are therefore NL/π states per unit interval of k-space. In the free-electron model, $E=\hbar^2k^2/2m$ and $dE/dk=\hbar^2k/m$, and so $\rho(E)=(NL/\pi)/(\hbar^2k/m)=(NL/\pi\hbar)(m/2E)^{1/2}$.

(ii) This time there is one allowed wave vector in the area $4\pi^2/A$ of $\boldsymbol{k}$-space, so that there are $A/2\pi^2$ states per unit area. For $|\boldsymbol{k}|$ between k and $k+dk$ there are therefore $(A/2\pi^2)2\pi k\,dk$ states having all the same energy. So $\rho(E)=(Ak/\pi)/(h^2k/m)$.

(iii) There are $V/4\pi^3$ states per unit volume. For $|\boldsymbol{k}|$ between k and $k+dk$ there are $(V/4\pi^3)4\pi k^2\,dk$ states of the same energy. So $\rho(E)=(Vk^2/\pi^2)/(\hbar^2k/m)$.

11.2 The stationary-state wave functions are $e^{i(k_1x+k_2y)}\sin q\pi x/a$, with $E=\hbar^2(k_1^2+k_2^2)/2m+\hbar^2q^2\pi^2/2ma^2$. The allowed values of k_1 and k_2 are as in part (ii) of the previous problem, while $q=1, 2, 3, \ldots$. Take first $q=1$. Then $E\geqslant\hbar^2\pi^2/2ma^2$ and the contribution to $\rho(E)$ is $Am/\pi\hbar^2$ as

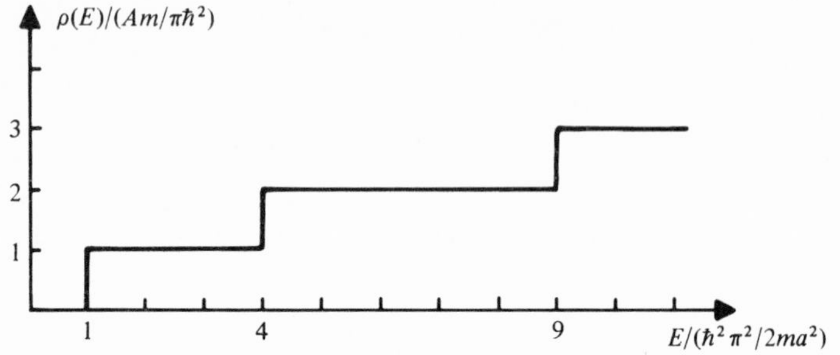

in part (ii) above. For $q = 2$, there is a similar, additional contribution; this begins at $E = 4\hbar^2\pi^2/2ma^2$. And so on.

11.3 From part (iii) of problem 11.1, for unit volume of the metal $\rho(E) = (2m^3E/\pi^4\hbar^6)^{1/2}$. The Fermi–Dirac distribution (11.21) gives the number of electrons $n(E)$ in the unit volume per unit E. Since $E = p^2/2m$, the number per unit p is obtained by multiplying this by $\mathrm{d}E/\mathrm{d}p = p/m$ and is

$$(p^2\,\mathrm{d}p/\pi^2\hbar^3)[\mathrm{e}^{(p^2/2m-\zeta)/k_BT}+1]^{-1}.$$

The momentum distribution is spherically symmetric in the free-electron model, so the number of electrons per unit volume of $\boldsymbol{p}$-space is obtained by dividing by $4\pi p^2$. Only electrons having $p_x^2/2m > V_0$ can escape, and for these $p^2/2m - \zeta \gg k_BT$, so that the exponential in the denominator dominates. The current is therefore

$$\int_{(2mV_0)^{1/2}}^{\infty}\mathrm{d}p_x\int_{-\infty}^{\infty}\mathrm{d}p_y\int_{-\infty}^{\infty}\mathrm{d}p_z\frac{ep_x}{m}\frac{\mathrm{e}^{\zeta/k_BT}}{4\pi^3\hbar^3}\mathrm{e}^{-p_x^2/2mk_BT}\,\mathrm{e}^{-p_y^2/2mk_BT}\,\mathrm{e}^{-p_z^2/2mk_BT}.$$

11.4 Near the valence band, $\rho(E) = N\delta(E+\Delta)$. The number of electrons in the valence band is obtained by inserting this into the Fermi–Dirac distribution (11.21) and integrating around $E = -\Delta$:

$$\frac{N}{\mathrm{e}^{-(\Delta+\zeta)/k_BT}+1} = N - n.$$

This results in the given expression for n. But also

$$n = \int_0^\infty \frac{A\sqrt{E}\,\mathrm{d}E}{\mathrm{e}^{(E-\zeta)/k_BT}+1} \approx A\int_0^\infty \mathrm{d}E\sqrt{E}\,\mathrm{e}^{-(E-\zeta)/k_BT}$$
$$= \tfrac{1}{2}A(k_BT)^{3/2}\pi^{1/2}\,\mathrm{e}^{\zeta/k_BT}.$$

Equating the two expressions for n gives $\mathrm{e}^{\zeta/k_BT} + \mathrm{e}^{(\Delta+2\zeta)/k_BT} = (2N/A)(k_BT)^{-3/2}\pi^{-1/2}$. When $\mathrm{e}^{-\zeta/k_BT} \ll 1$, the first term is negligible and the result follows.

12.1 The function $u_k(x)$ determined in problem 10.3 may be used to construct the solution given in (12.1) and (12.2). The constants R and T are determined by making ψ continuous at $x = 0$, and disc $\psi' = -U_2\psi$ there. The same level structure obtains as in problem 10.3.

Consider now the trapped-electron levels. In $0<x<L$, where the potential vanishes, we have $\psi(x)=C\,e^{\alpha x}+D\,e^{-\alpha x}$, where $E=-\hbar^2\alpha^2/2m$. Writing ψ as in (12.4), and making $v_1(x)$ periodic, we find that in $L<x<2L$ we have $\psi(x)=C\,e^{\alpha(x-L)+\kappa x}+D\,e^{-\alpha(x-L)+\kappa x}$. Imposing the conditions at $x=L$, that ψ is continuous and disc $\psi'=-U_1\psi$, gives two equations. With the even-parity solution, one of the two corresponding conditions is satisfied automatically. The other, together with the two previous equations, gives on eliminating C, D and $e^{\kappa L}$

$$(U_1U_2+\tfrac{1}{2}U_2^2-2\alpha^2)\tanh\alpha L=2\alpha U_1.$$

There is no negative-parity solution.

12.2 In the case where there is no external potential, the numbers of electrons passing across the interface in each direction are equal. Because the valence band of the p-type material overlaps the conduction band of the n-type region, some of the electron flow is the result of tunnelling between the two; the remainder is due to electrons passing directly from one conduction band to the other, as in the normal junction diode. When the n-type material is connected to a negative potential, so that its energy levels are shifted upwards relative to those in the p-type region, the flow of electrons from p to n is reduced. But the tunnelling of electrons in the other direction is greatly increased, because of the large number of empty electron states in the valence band of the n-region. The current is then greater than for a normal diode, in which the tunnelling does not occur. When the potential difference is increased to such a value that overlap of the p-type valence band with the n-type conduction band no longer occurs, the tunnelling ceases and the situation is the same as for the normal junction diode (see figure 12.5).

(*a*)

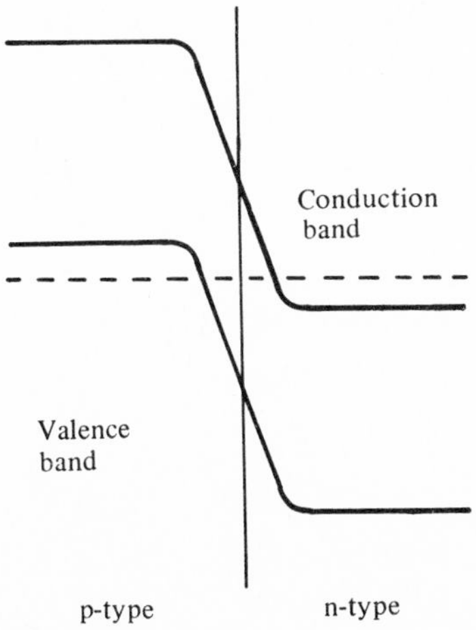

(*b*)

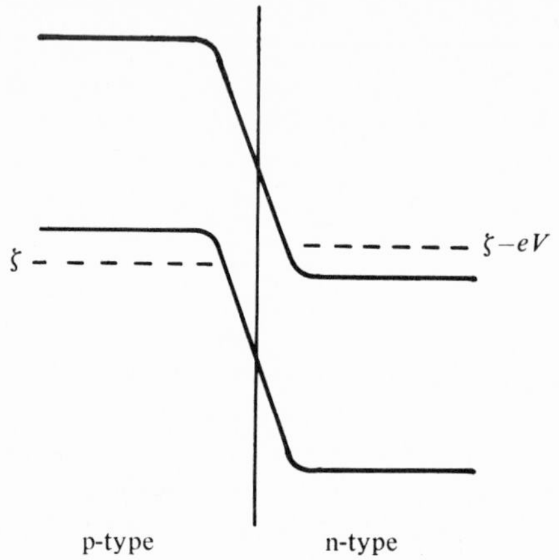
ζ
ζ − eV
p-type
n-type

(*c*)

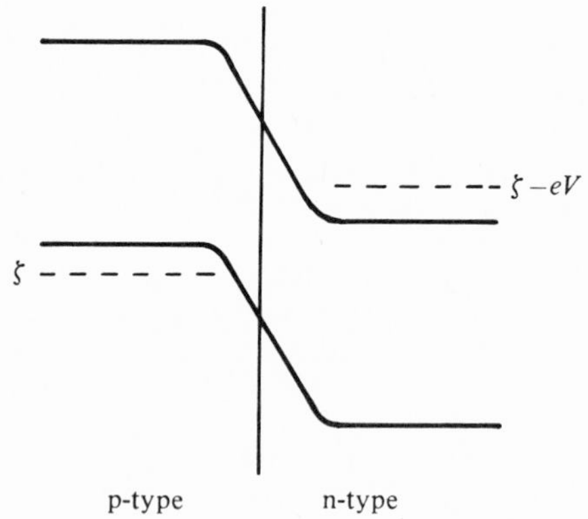
ζ
ζ − eV
p-type
n-type

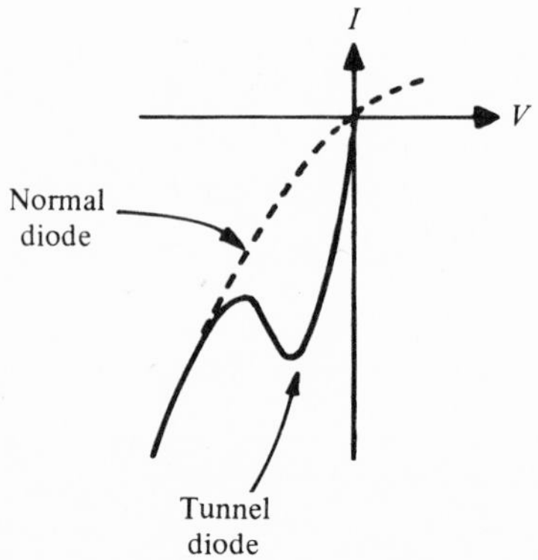
I
V
Normal
diode
Tunnel
diode

INDEX